Quantum Consciousness

Norbert Wrobel, living in Berlin, studied medicine and approved in 1984 as a doctor. In a wide-ranging academic basic training at the Free University of Berlin, he subsequently specialized in the fields below internal medicine, intensive care and emergency medicine and later in geriatrics, and has since then been active in the in-patient care. Because of changes in society that produces more and more elderly people, health professionals are constitutively faced with new, unfamiliar and complex situations. This subject, however, to this day, is an out-dated mechanistic-physical way of thinking that has developed more than a hundred years ago. Norbert Wrobel has therefore made himself up to move away from this way of thinking and to explore, what actually holds the world together in its innermost.

The mathematician *Klaus-Dieter Sedlacek*, born in 1948, has lived in southern Germany since his childhood. He studied mathematics and computer sciences in addition with physics. After graduating in 1975, and some years of professional experience, he founded his own company which dealt with the development of application software. He run his business for more than twenty-five years. In the second half of his life he has devoted himself to his private research projects. He has set himself the task to investigate deeper the physics of information, meaning and consciousness, in order to reach a wider audience. In 2008 he published a startling fiction book entitled: »Immortal consciousness - spacetime phenomena, evidence and visions«.

Norbert Wrobel
Klaus-Dieter Sedlacek

Quantum Consciousness

Natural foundations of a theory of evolutionary quantum consciousness

Research Library

Bibliographic information at the German National Library:
This publication has been recorded by the German National Library in the German National Bibliography; detailed bibliographic data can be found on the Internet at http://dnb.d-nb.de.

Original edition

qrdrg.xonl.de

Production and publishing house:

Herstellung und Verlag:
BoD – Books on Demand, Norderstedt
ISBN 978-3-7386-2750-3

Contents

0. Preface of the physician Norbert Wrobel

The interdisciplinary dialogue with Klaus-Dieter Sedlacek, started on the question: „What is disease?", initially led to a thorough debate in terms of: quantum physics and, in particular, to questions of quantum information theory. This dialogue, held very impassionately, has helped me since to let go the „old" mechanistic way of thinking and, at the same time, helped me to turn towards the wider concept of „new" physics. I am convinced now that „modern man" is regarded as a self-organized, dissipative non-equilibrium-system and that it operates on the same principles as the 4-dimensional world in which we live: by means of elementary information and mere chance (see: booklet "Leben aus Quantenstaub - Life out of quantum dust").

In this script, I will now undergo a differentiated implementation of the two links: 1) the metric-free vacuum[1] and 2) the real world, namely consciousness, triggered by processing information. Usually, the term „consciousness" is associated with higher, cognitive performance. In the course of this dialogue however we referred to by our chosen syntax and semantics, consciousness was assigned to in accordance with information-processing, as a principle of quantum physics.

Everything, that exists in this world, that is all quantum objects, are in principle integrated into this process. Information, fluctuation and decoherence, entanglement and evolution included, have congruently been regarded as elementary. With these ingredients it is possible to produce a matter-containing reality in a 4-dimensional world.

1 A metric free vacuum as a non-local physical field (not belonging to the space-time) contains nothing except the information that is equivalent to energy or matter

Consciousness is associated with something alive, imminently sensed by someone´s life-experience. It is intuitively associated with organic matter. For evolutionary reasons we believe that human beings show the highest developed consciousness, due to a long preceding onto- and phylogenetic process.

There is however evidence in the geological history of our Earth, that inorganic matter existed before living matter. If, from a quantum-theoretical point of view, a kind of spiritual interaction of all quantum objects with the metric-free vacuum is possible then the question compulsorily arises, whether not non-organic matter must have consciousness, too. If it has, all information-processing can principally form the world we live in, triggered by all quantum objects.

> *"All the world is a stage, we are merely players, performers and portrayers, each another's audience outside the gilded cage!" (From: RUSH 'Limelight' in "Moving pictures")*

We warmly would like to say thanks to Christian Wrobel for his passionate translation.

Berlin in autumn 2014

Norbert Wrobel

1. Preface of the mathematician Klaus-Dieter Sedlacek

It always amazes me what synergies an interdisciplinary dialogue can develop. The first guided, e-mail conducted dialogue with Norbert Wrobel led to the publication of „Leben aus Quantenstaub - Life out of Quantum Dust, engl.“ and, among other things, to the knowledge, where the excess of positive energy or material in the universe originates, a finding that mainstream physics can not explain.

To put it more popular, the excess energy that really should not exist according to the law of conservation of energy, does therefore exist in our universe, because it is „borrowed“ energy that comes from vacuum fluctuations, whose re-payment or compensation is a long time coming until the end of time.

In the new dialogue, there are again exciting results for the reader. Even though I am not at home in the department of microbiology myself, I was able, in co-operation with Norbert Wrobel, to discover something new which a microbiologist, who is specialized exclusively in his faculty, might not have seen. The more I read about the behaviour and life of microbes, the more I was fascinated by microbiology, and questions arose that have obviously not been answered by microbiologists, yet.

Take, for example, the so-called moon-milk, researchers had discovered in the depths of a tyrolean ice cave. Each drop of this moon-milk contains a huge amount of microbes living in a community. It is hard to imagine that, under the extreme dark and frosty conditions of the cave, the members are able to survive and to multiply. Well, some of the microbes may serve as food for others. There is a beginning of the food chain, though. And: how do microbes, at the beginning of the food chain survive and replicate?

Where does the energy for their metabolism come from? Which kind of information- processing ensures their survival? And finally a question, the answering of which I am particularly interested in: might information-processing fulfil criteria of a rudimentary form of consciousness, which is necessary for their survival? If so, this would have far-reaching implications for our understanding of the evolution of life. I think we have found answers, and Norbert Wrobel even formed a new theory, that now everybody can refer to.

Spain in autumn 2014

Klaus-Dieter Sedlacek

2. Quantum-associated aspects of the development of life (Phylogenesis)

2.1 Consciousness as onto- and phylogenetic development processes

@ Dear Mr. Sedlacek (following KDS):

... the term "consciousness" is usually associated with cognitive brain functions. In the dialogue we had so far [2], we directly assigned consciousness to information-processing, according to the way we speak and think (syntax, semantics). According to DIN IEC 60050-351, information-processing provides the connection between the metric-free vacuum and the real world[3]. Everything in this world, that is, all quantum objects, are principally involved in this process. We have agreed that

- Information
- Fluctuation[4] (=mere chance) and decoherence[5]

2 Wrobel, Sedlacek (2014) Life out of quantum dust - Elementary information and mere chance in the middle of nothing as elements of a 4-dimensional quantum world. German: Leben aus Quantenstaub.

3 On the concept of consciousness be aware of the box on p. 66.

4 **Fluctuation** referred to a random quantum physical process that effects a change of circumstances and conditions.

5 **Decoherence** is a decay of the state of a quantum system when completed so far, if it occurs with its environment, whereby both, the state of the environment and the state of the system, is changed

- Quantum mechanical entanglement[6]
- Evolution

is regarded as elementary[7]. With these ingredients it is possible to generate matter-containing-reality within a 4-dimensional world.

From one´s own life experience, we connect consciousness with something alive and assign it intuitively to organic matter. For evolutionary reasons we next assume for humans that they show the highest level of development, speaking in terms of consciousness.

If this is indeed the case, a long ontogenetic and phylogenetic development process (Haeckel, Darwin) must have preceded. Aprioristic organic consciousness (Kant), on the other hand, can probably be ruled out, as the history of development of our planet Earth has shown that inorganic matter came before life.

On the basis of the unity of nature, consciousness is understood as a universal property of all "beings" within the 4-dim world, which consequently means that an inorganic substance must have consciousness features, too.

How could one now approach the problem from a scientific perspective, to find a solution for something, which totally contradicts to our experience of life? Should even a boulder have consciousness, too?

---

irreversibly.

6 **Quantum entanglement** is a physical phenomenon in which two or more particles enter a non-local connection to each other. Measurements of certain physical quantities of entangled particles are correlated. This means, if we measure the quantum particle properties at particle A (i.e. spin), then the correlated property (negative spin) is also found without delay (instantaneously) in particle B.

7 Wrobel, Sedlacek (2014): *Quantenstaub*, p 93

A much promising way is to look for evidence-based knowledge from natural science: great naturalists such as Ernst Haeckel[8], who

> *„... demonstrated a special way in the battle of world-views for the individual honest reader, who struggles for pure reason-based-knowledge. To his firm conviction, this alone would lead to the truth as one possibility: empirical natural science and the hence founded monistic philosophy"* [9].

With respect to another work by Haeckel, the "Natürliche Schöpfungsgeschichte " (History of natural creation, engl.), Darwin[10] in full recognition said:

> *"... if this work had appeared before my essay (The Descent of Man) had been written, I should probably never have completed it. Almost all the conclusions to which I have arrived I find confirmed by this naturalist, whose knowledge on many points is much fuller than mine "* [11].

In: "Die Welträthsel – Wonders, engl.", Haeckel now raises the issue on consciousness, especially ontogenetic and phylogenetic development aspects respectively, and in the psychological part of "Die Seele – The Soul, engl." he carries out among other things:

> ***"Soul is a sum of vital phenomena**, which are tied to a specific material substratum, which I shall call **psychoplasm**, because this substrate has been detected as*

8 Ernst **Haeckel** (1834-1919), Zoologist and philosopher, is considered to be the "German Darwin".

9 Haeckel: Introduction to "The Welträthsel" http://www.zum.de/Stueber/haeckel/weltraethsel/weltraethsel.html

10 Charles **Darwin** (1809-1882), British naturalist, published his theory of evolution at the age of 51.

11 Charles Darwin: Introduction "The Decent of Man"

> *belonging to the group of plasma body. Even* ***for atoms the simplest form of sensation and will is inherent****, too*[12]*, as a most primitive type of a universal "soul".*

Following his "Theory of Consciousness", he hypothesized on conscious and unconscious mental life. Consciousness is, as any other mental activity, a natural phenomenon and therefore subject to the law of substance. A distinction is made between the awareness of the outside world as the "world consciousness", and an inside mirroring of all our ideas, feelings and strivings or volition, known as "self-consciousness".

He assumed that the earliest human ancestors, unicellular **protozoa** - from the systematic unity of protists such as the amoeba - are only "inspired", without consciousness, and brought this property in connection with special plasma molecules (**plastidules**):

> *"The mental processes there are thus the bridge which connects the chemical processes in inorganic nature with the mental life of animals*[13]*".*

Due to similar elementary structures, such as cell and plasma, he concluded that **even plants must be animated**.

In his molecular genetic theory, he formulated a tentative hypothesis, according to which the original development processes emerged strictly mechanically, and from physical-chemical elements:

---

12 A modern experiment that seems to confirm this statement is the double-slit experiment with elementary particles. On his way through the column, the elementary particle recognizes (= form of sensation), whether one or two gaps are opened. If two slits are open, it chooses a path that can not be predetermined by one or both columns (= will).

13 Among the **protists**, all unicellar to morecellular eukaryotes are included, i.e. algae, protozoa and some fungi.

„Die Perigenisis der Plastidule“ (Perigenisis of plastidules, engl.), or the “Wellenzeugung der Lebenstheilchen“ (Wave-procreation of the vital particles, engl.).

He assumed that:

- **protoplasma**[14] is the physical basis of life,
- the simplest forms of life such as Protogenes primordialis (Protomonas) have homogeneous, structureless protoplasm - similar and homogeneous as is a crystal, and stand on the borderline of organic and inorganic that is of the so-called animate and inanimate nature,
- life is also bound to formless substances of certain physical properties and chemical composition,
- properties of the organic is worthy to inorganic matter, too, and thus common to all natural bodies , and - ultimately,. of the atoms,
- each atom has an inherent sum of force and is in this sense "inspired",
- without the assumption of an "atomic-soul", the phenomena of chemistry or physics, such as attraction and repulsion, or motion of the atoms, are inexplicable in the sense of **feeling and will**,
- each ground atom is equipped with a constant and eternal soul is therefore immortal,
- only the countless and ever-changing compounds of atoms are mortal, though.

14 **Protoplasm** nowadays is a rarely used common name, filling the cells with plasma-like basic structure, the cell nucleus included. The cytoplasm without nucleus and without structurally distinguishable areas of special function is nowadays denoted cytoplasm.

- plasma molecules (plastidules) have a **memory** at their disposal - which is essential for reproduction, and can be described as vital particles,
- branched undulations of plastidules are regarded as mechanical cause of the biogenetic process (= periodic mass movement)[15].

In his extensive elaboration "Generelle Morphologie - General morphology, engl." (Haeckel, 1866) and later in his finely profiled "Natürliche Schöpfungsgeschichte - History of natural creation, engl." (Haeckel, 1868), he described **cellular psychology** (cellular theory of consciousness) as a life characteristic of each cell. In contrast, in his "atomistic theory of consciousness", according to which **each chemical element would have elementary consciousness**, he especially featured "unconscious mind" characteristic to inorganic-inanimate.

Under no conditions he would allow transcendent explanations for supernatural phenomena of certain brain functions:

> *"The neurological problem of consciousness is only a special case of the universal cosmological problem, namely the substance question".*

Methodologically, however, he accepted an introspective approach as subjective, inner method, according to which the immediate certainty of the ego should emerge as "self-consciousness": "Cogito, ergo sum - I think, therefore I am" (Descartes).

With regard to the wealth of Haeckel's elaborations, a set of questions arises quite inevitably:

- Are the nearly 150-years-old findings transferable to

15 Ernst Haeckel: *Die Perigenesis der Plastidule* (1876)

the present day at all?

- Can these findings, mainly from the then usual protists classification (primeval being, first fruits) - a group of related unspecified microscopic organisms such as algae, protozoa or some few fungi – that have been derived as a systematic unity, be transferred to today's living-being-classification with the domains bacteria, arachaea[16] and eukaryotes?
- How to verify some sort of consciousness in inorganic-inanimate at all?

The vacuum theory[17] conducts fundamental physical information processes, which take place within the metric-free vacuum. Comparable with a store, in which all processes (interactions) from space-time[18] are recorded and never deleted. Information-processing there would fulfill the criteria for consciousness. Therefore, they can also be described as units of consciousness. The consciousness units consist of an information type which is equivalent to energy or mass[19].

@@@

Terms

- A **metric-free vacuum** is a non-local physical field (i.e. not belonging to the space-time), containing nothing but information, being

16 Archaea are single-celled organisms

17 Sedlacek (2012) *Der Widerhall des Urknalls* - Echo of the Big Bang, p 175ff

18 In the theory of relativity, the **spacetime** designates the unification of space and time in a single four-dimensional structure, with special properties. The space-time serves as a model for the geometry of the real world.

19 ibid p. 176f

equivalent to energy, or matter

- **Quantum - biology** is the denomination given to a branch of bio-physics. It deals with the influence of quantum effects on living cells of an organism. Energetic processes and changes that may be in the range of atoms and molecules, are examined[20].

- As **plasmon (physics)**, referred to as the quantized variations in the charge carrier density in semiconductors and metals; quantum-mechanically they are treated as quasi-particles (= excitation in the many-body system). The term is a common abbreviation for plasma quantum oscillation. What the photon represents for electromagnetic waves, is the plasmon for oscillations for the fermigas[21] of metals[22].

- **Plasmon (biology)** means: 1) physical plasmon, the physical totality of extrachromosomal, plasmatic hereditary factors of a cell or organism. For the physical plasmon chondroma, the (physical) plastome (in plants) as well as the genes of other plasma factors are counted. 2) typical plasmon, the totality of the types of extrachromosomal genes of a cell, e.g., the entirety of their typical chondroma and plastome; the joint features that make two cells or individuals in terms of extrachromosomal genetic factors, genetically identical.

20 see http://de.inforapid.org/index.php?search=Quantenbiologie

21 System of identical particles of type fermions in quantum physics.

22 see http://cms.uni-konstanz.de/index.php? eID = tx_nawsecuredl & u = 0 & file = file admin / physics / Ganteför / pdf files / SS05 / 08.pdf

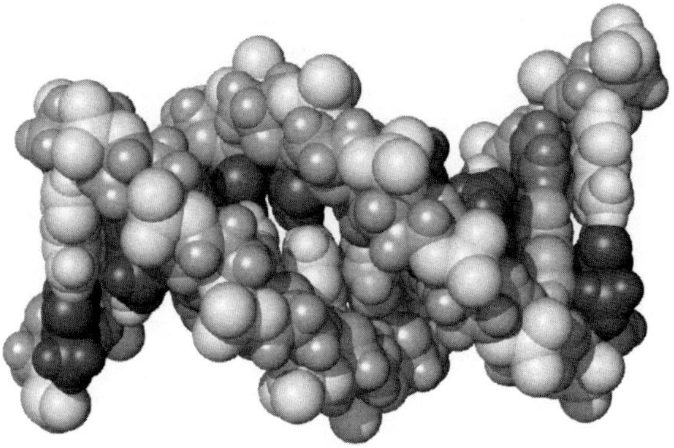

Fig. 1: DNA portion ; CC0 by Yikrazuul

- **Genome**

1) physical genome, which is contained in a virus, a single cell, or in the cells of a multicellular organism, as a physical totality of genes and genetic signal structures, as well as the DNA regions (deoxyribonucleic acids), to which no, or not yet, function can be allocated. Thus, the entire physical genome of an adult human contains about 20,000 to 25,000 genes. Since no two cells or individuals may share the same specific DNA, they can not have the same physical genome. In this sense, two cells only have the same genome, if they possess genes of the same type, i.e., in which their DNA has the same sequences, or functions. For this reason, the physical genome must be distinguished from the typical genome

2) Typical genome, which, in a virus, a single cell, or in the cells of a multicellular organism, entirely

contains all varieties, or types of genes, and genetic signal structures. The cataloguing of this typical human genome is, among others, one main target of the Genome Project.

Organisms (eukaryotes), all constructed on nucleated cells (eucyte), are divided into both, physical and typical genome in the nuclear genome, that is contained in the chromosomes of the cell nucleus DNA, which is often also referred to as the real genome. Further, in the cytoplasmic genome contained organelles[23] (plasmon). It is a result from the genome (mitochondrial genome or chondroma), localized in mitochondria. Additionally, in plants and green algae from the genome (plastid genome or plastome), localized in plastids[24]. Again, in each case it must be distinguished between the physical typical nuclear genome and plasmon, chondroma, or plastome.

- **Chondroma**

1) Physical chondroma, which is the physical totality of DNA of a mitochondria[25], or even all mitochondria of a cell, or an organism (extrachromosomal genetic factors). The proportion of mitochondrial DNA (mtDNA, mt-DNA) to the total DNA of a cell ranges between 0.5% and 20%.

2) Typical chondroma, the entirety of the types of

23 An **organelle** is a structurally definable area of a cell with a specific role.

24 Cell organelles of plants and algae that have emerged from endosymbiotic living cells. They are required, inter alia, for photosynthesis.

25 **Mitochondria** is a cellular component that enables the cell respiration and hence the energy supply of a eucaryotic cell. See also Fig. 2 on page 29 and the box on page 28.

mitochondrial genes, and other functional units of the mitochondrial DNA.

- **Plastome**

1) physical plastome, the physical totality of the DNA, or genes and other functional units of a plastid, or plastids of a cell or all of a plant.

2) typical plastome, typical plastid genome, all of the types of genes and other functional units of a plastid; the set of plastid DNA, which can occur multiple times within a plastid (polyploidy), or in terms of the two plastids, two cells or two plants, which can be genetically identical.

- **Signal structures** are acting as signals in the recording or disclosure of information structures. Biological signal structures are particularly known at the molecular level (signal sequences), e.g., the origin of replication for signalling of the start of DNA replication.

- **Plasma factors**: collective name for the genetic information-bearing units inside the cell´s plasm (=cytoplasm), such as mitochondria, chloroplasts, and kappa factors in eukaryotes cells, or plasmids in prokaryotic cells. The encoded characteristics by plasma factors in eukaryotes cells are not in accordance with the rules of Mendelian inheritance (cytoplasmic inheritance).

2.2 Biology and physics of the organic and inorganic

@ KDS:

... I meanwhile found out in biology books from the 1950s a definition of plasmon (biology). Into these biology books important findings have been incorporated since the 1920s. Accordingly, it is an umbrella term for a specific cytoplasmic region including the extrachromosomal inheritance. In contrast, the genome indicates the chromosomal and extrachromosomal inheritance in their entirety[26].

The terms "signal structures" and "plasma factors" seem very interesting to me, with which an analogy may be made with the plasmon in physics[27].

The plasmon (biology) - in any case, is clearly responsible for the extrachromosomal inheritance; that means, Mendel's laws do not apply to this type of inheritance. Possibly - occurring in biology, the signal structures and plasma factors are involved in the triggering of **information-processing.** In contrast, the plasmon (physics) in quantum physics is treated as "quasiparticle".

Imagine now the question, whether plasmons in biology or physics have effects according to the same principles: in physics there is a concept of surface-plasmon-resonance (SPR). For example, there are gold nanoparticles, but also diamonds in nano size are used. The there detected quantum effects are apparently mediated by "quasi-particles". Could you clearly explain what is meant by this?

Could any cross-connection of quasi-particles with the extrachromosomal inheritance be made (plasmon biology, signal structures, plasma factors)?

26 **Oehlkers** (1956); Hagemann; Dessauer (1954)
27 see http://de.m.wikipedia.org/wiki/Plasmon_(Physik)

The influence of light (photons) on the growth of plants is sufficiently well established. These are likely to plastids (chromoplasts), playing a prominent role. In sea-weed, quantum effects have already been clearly established[28].

Thus, for example, algae in the conversion of light into energy operate very efficiently. How efficient they actually are can be estimated, because researchers are extremely happy with the technical photovoltaic efficiencies up to 12%[29].

In contrast, the yield in natural photosynthesis is at 95% (!). How can we explain these big differences?

@@@

Physicists open the doors for quasi-particles

Innsbruck physicists have created quasi-particles in a quantum system, while first observed the propagation of quantum entanglement in a many-body system.

More at:

http://www.juraforum.de/wissenschaft/physiker-oeffnen-tuer-in-die-welt-der-quasiteilchen-485617

http://derstandard.at/2000002874314/Physiker-studierenerstmals-Informationsfluss-in-groesserem-Quantensystem

http://www.oe-journal.at/index_up.htm?http://www.oejournal.at/Aktuelles/!2014/0714/W2/31007oeaw.htm

28 see http://www.spektrum.de/news/alge-nutzt-quanteneffekte/1020976

29 see http://www.prophysik.de/details/news/6426371/Druckfaehige_Perowskit- Solarzellen_immer_stabiler.htm

@ Dear Mr. Wrobel (following NW):

... in the first post (Juraforum) the author seems to apply little understanding of modern physics, although the contribution is not objectionable as to its contents. Like in the two other contributions there is concern to demonstrate, among other reasons, whether quantum effects - here entanglement, occur not only in the case of photon pairs, or pairs of other quantum particles, but also in large systems, such as in large and warm solids - if not even in living organisms, thus everywhere in the world around us. For this purpose, an experiment at the University College was carried out already in 2003:

A piece of lithium fluoride is exposed to a magnetic field. This salt is magnetizable. The atoms in the salt behave like small magnets and are eager to align along the magnetic field lines. The researchers measured, how fast the atoms lined up. Obviously, this happened much faster than could be explained by classical interactions between the atoms, hence only original entanglement is eligible.

The atoms of a salt must therefore be somehow entangled. In more recent experiments, as described in the articles above, there is also concern, how fast correlated quantum systems in larger states spread up in order to conclude again on entanglement. Eventually it will probably be possible to prove scientifically correct - based on the results, where, how, and to what extent, entangled states occur in our everyday world, and also in humans.

@@@

@ KDS:

... how could the quasi-particle experiment be interpreted in our speech and our way of thinking (syntax, semantic)? There must be a process and "borrowed" energy, respectively. Interestingly, the Innsbruck physicists want to quantify and standardize entanglement.

You already formulated solutions as part of your vacuum theory. Does the approach, chosen by the researchers, make sense at all? If so, how could one come up with meaningful questions that fit to your solution?

@@@

@ NW:

... I will gladly try to use the intellectual scalpel to dissect the reports of the quasi-particles in order to investigate, what the whole thing could be good for.

Under quasi-particles physics understand a wave-like excitation (oscillation) of a many-body system or solid, which **can be treated mathematically** as a particle in quantum physics.

A quasi-particle is then neither the solid, nor the amount of particles, which is put into vibration by excitation, but it is the excitation, or it is the vibration itself. A quasi-particle is - due to its definition, nothing that belongs to the real world, it is quite simply a mathematical construct, a square or a formula, for example. Which features in quasi-particle are really real, physics have not told us, yet. Regrettably, physicists love to take again and again their abstract constructs at face value, and hence, to develop real entities.

Only the many-body system is real, or the solid body, but not the quasi-particle.

To make quasi-particles at all, excitation is required. That is, the many-body system, or the solid must be externally supplied by energy, e.g., by a laser beam. Without this supply of energy there is no stimulation, and thus no quasi-particles.

Now, what is the purpose to produce quasi-particles? Only to ascribe them properties of real particles such as mass, momentum, energy, wavelength and spin[30]. This means, you can measure something, you can describe their behaviour. Scientists can explore quantum phenomena experimentally with real particles, which are otherwise not accessible. Quote from the report:

> *"It is also a big goal to make use of quasi-particles for quantum information- processing". But the role of quantum physics in transport processes - as they occur similarly in biology, could be studied on this platform.*

For quasi-particles there are obviously only vague ideas for use in quantum information- processing. As far as the transport processes in biology is concerned, especially during photosynthesis under the influence of quantum phenomena, we have already found rated articles in the internet. I can well imagine that in this field, with the concept of the excited many-body systems (quasi-particles), there will be further research in the near future.

Speculatively, I would like to note in this context: it is known that in the human body signals are mainly transmitted

30 **Spin** is the intrinsic angular momentum of particles. Clearly, however, physically incorrect, you can explain the spin by the swirl of a compass needle.

electrically by stimulations of the nerves. The work of the brain is therefore very well observable from the outside, because thought processes are accompanied by measurable electromagnetic signals and waves.

But this does not explain the high speed of thought processes, which require an interaction of distant brain areas. The spaced areas of the brain are in fact not generally associated with those high number of necessary nerve pathways to explain the speed, according to the classical point of view. Here you have probably to assume that there are quantum physics correlations between brain areas. That is the information is transported with the aid of quantum entanglement. Since the brain, or the nerves, can be seen as a many-body system, I assume that, from the concept of quasi-particles experimental results, scientific conclusions are possible.

How should we now interpret quasi-particles in our speech and way of thinking (syntax, semantic)? The "stimulus" that led to the concept of quasi-particle is understood as a real process. It includes the abstract information which distinguishes this particular stimulation to other stimulations. Processes associated with abstract information have the character of structural information and are thus equivalent to energy. Energy, however, is not borrowed from the metric-free vacuum, but it comes directly from the 4-dimensional world and is supplied in this example by laser light.

What is the "nutrition-value" of quasi-particles concerning our more philosophical reflection on the foundations of the world? I think, as good as any other. It is not scientific knowledge that has to somehow fit into a theory, such as the wired behaviour of quanta, or the amazing phenomenon of "spooky action at a distance", i.e. quantum entanglement.

Quasi-particles have more the character of an experimental set-up, such as the double-slit belongs to the double-slit experiment. The double-slit is not the scientific result. The scientific result is the amazing interference of photons or electrons on the observation screen. To which conclusions the concept of quasi-particles will lead us, we do not really know. We can assume it only vaguely.

@@@

Terms

- **Mitochondria** are considered as power plants of a cell. Originally, a mitochondrium was a small bacterium that was able to gain energy by respiration of oxygen.
- **Chondroma** represents the extrachromosomal fraction of the genome of a cell and is found in the mitochondrium. It has been specialized in the energy-producing task.
- **Plastids / chloroplasts** are the photosynthetic factories of plant cells. Originally, plastids derived from the protoplasm of blue-green algae (= cyanobacteria) or other prokaryotes.
- The **plastome** represents the extrachromosomal fraction of the genome of a plant cell and is assigned to the plastids. Plastids are required for gaining energy from photosynthesis. They are also able to produce oxygen.

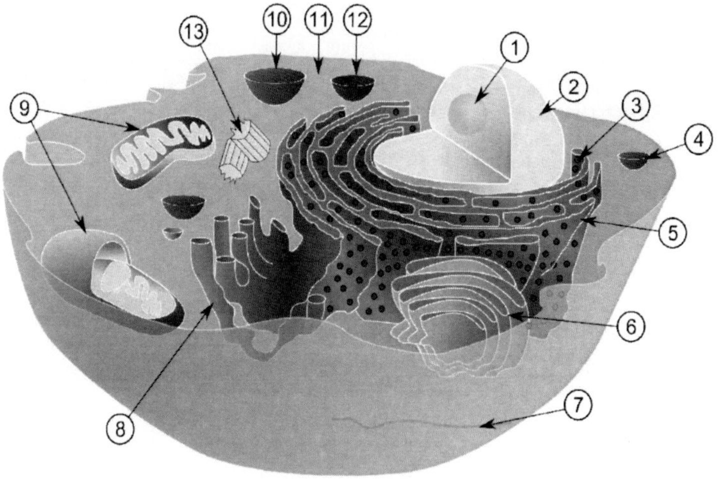

Fig. 2: Organization of a typical eukaryotic cell 1. nucleolus (chromosomal DNA) 2. nuclear envelope 3. ribosomes (duties of protein synthesis) 4. vesicles transporting materials 5. Rough endoplasmic reticulum (ER) separated from each other cellular processes 6. Golgi apparatus (tasks of cell metabolism) 7. microtubules (eg for movement and transport within the cell) 8. Smooth ER 9 mitochondria (= cell organelles with their own genetic material. tasks for cellular respiration and energy supply) 10. chloroplast (photosynthetically active in green plants). In addition, plant cells have a vacuole (not shown) as a digestive organ 11. cytoplasm (= the cell fills basic structure) 12. peroxisome (detoxification apparatus) 13 centrioles (transport and support tasks).

2.3 Endosymbiotic theory

As researchers want animals to teach photosynthesis

Only plants use solar energy? Not at all, some animals have found ways to "being satisfied" while sunbathing. Researchers want to teach this to other living beings - and in the future maybe humans, too.

More at: : http://www.tagesspiegel.de/wissen/ungewoehnliche-tiere-die-lichtfresser/10188286.html

@ KDS:

... we should come back to the question, how animate beings including algae, interact through information-processing with the metric-free vacuum. I discovered a very interesting article today. Accordingly, the slug Elysia timida gnaws a hole in a thread algae and then sucks out the cell contents, chloroplasts (plastid belonging) included. The absorbed food is not digested, but deposited along the digestive system. The highlight: the chloroplasts, now in the snail, continue to work as before in the algae and can serve as its energy supplier. Elysia timida is thus an animal with built-in solar panels, a hybrid between animal and plant life.

In sea-weed quantum effects have already been demonstrated. I wonder whether - deduced from the special Elysia case through quantum effects, universal properties are being performed in living beings? In plastids (umbrella term **plastome** in plants for the extrachromosomal fraction of the genome), light is known to be „recyclable“. Could it also be similar in chondroma (another extrachromosomal proportion of the genome), which are assigned to mitochondria?

Mitochondria and plastids/chloroplasts of eukaryotic cells developed very early in evolution by endosymbiosis[31] **(Endosymbiotic theory**[32]). As they developed, prokaryotes, such as cyanobacteria and other - in case of plastids/chloroplasts or mitochondria-related a-proteobacteria - were added, but not digested, or broken down by primordial eukaryotes (predecessor of the animal or plant cells), and "kept alive" (= process of phagocytosis). In primordial eukaryotes, phagocytosed erstwhile prokaryotes propagated in interaction with them, but cellularly and functionally regressed over time. In further development, they instead took over specific functions from primordial eukaryotes. Now, mitochondria in eukaryotic cells multiply by growth and sprouting, plastids/chloroplasts, however, by fission[33].

The process of endosymbiosis is probably went along as follows:

About 2 billion years ago, when oxygen in the air had already accumulated through photosynthesis by blue-green algae (= cyanobacteria) ...

... bacteria (= anaerobic host cell) - with low, i.e. inefficient energy production by fermentation processes - phagocytosed smaller bacteria (= aerobic cell) with effective and efficient energy production by oxygen respiration, but without digesting them. These "new" bacteria, equipped with something that now is called mitochondria, was now able to feed heterotrophicly and are considered as precursors of heterotrophic eucaryots, unicellular protozoa and animals.

31 Cohabitation or socialization of individuals of two different kinds, in which one of the parties will be included in the body of the other.

32 The **endosymbiotic theory** was published for the first time by the botanist Andreas Schimper in 1883.

33 see http://www.pnas.org/content/95/8/4368.full

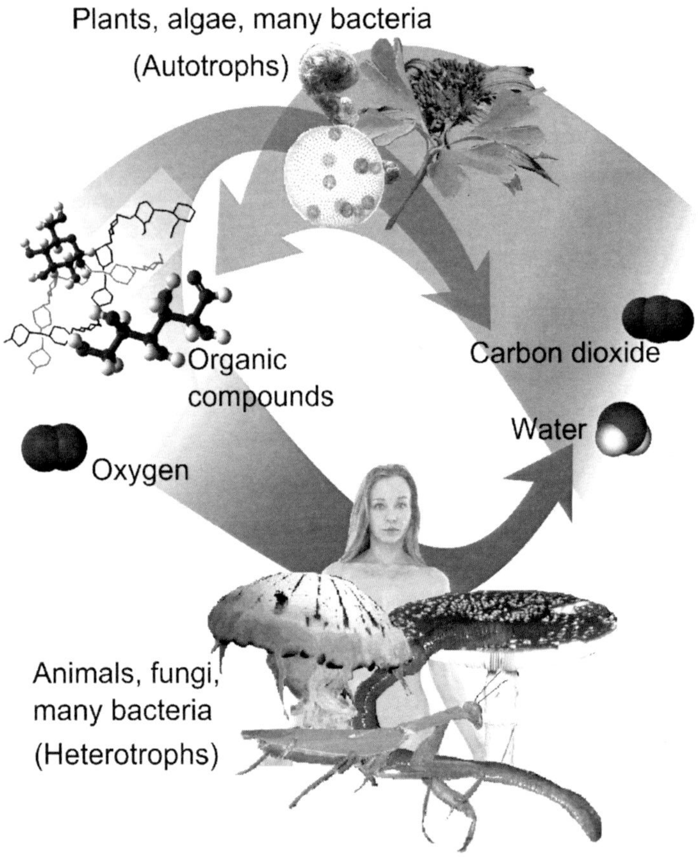

Fig. 3: Overview of cycle between autotrophs and heterotrophs. Photosynthesis is the main means by which plants, algae and many bacteria produce organic compounds and oxygen from carbon dioxide and water .
CC0 by Mikael Häggström

About 1.5 billion years ago, the previously described process was repeated by already heterotrophic bacteria, which now included ...

... blue-green algae (=cyanobacteria) and similar structures as endosymbionts with existing chlorophyll or chlorophyll analogies - with principal photosynthetic capabilities - and were therefore able to feed autotrophicly with the help of the now existing plastids/chloroplasts[34]. They are ancestors of autotrophic eukaryotes, unicellular protozoa and plants.

Geneticly, DNA-chondroma (mitochondria-related) and the DNA-plastome (plastid-related) specialized further by symbiosis processes within cells:

Photosynthetic pigments are chemical compounds which reflect only certain wavelengths of the visible light. This makes them appear coloured in the reflected wave fields. By creating abilities to absorb light at certain wavelengths, in nature many pigments play an important role as an energy interceptor. Photosynthesizing organisms - it is primarily these dyes, that absorb solar energy and make it usable. Since pigments quite typically absorb only certain wavelengths, mostly an ensemble of different dyes is found, so that the widest possible range of the solar spectrum can be exploited.

In our book: „*Leben aus Quantenstaub* - Life out of quantum dust, engl." you describe a class of electromagnetic waves, to which photons also belong[35]. When algae with their light- and quantum-associated plastome which developed later in evolution, started a symbiotic relationship

34 **Autotrophy** is the ability of living beings to build up their own building materials solely from inorganic materials. This fabric structure requires energy.

35 Wrobel, Sedlacek (2014) *Leben aus Quantenstaub* – Live out of quantum dust, engl, p. 75

with eukaryotic cells, chondroma (mitochondria-related) must have previously been part of an evolutionary principle.

Possibly equipped with skills to „exploit“ all kinds of electromagnetic waves. Now it would be highly interesting to find out, whether and how quantum processes affect in chondroma. Are you able to describe these processes by your theory?

2.4 Mode of action of plastids and mitochondria

@ NW:

... I will try to record - in a simplified form, what I know about the processes in cells and will draw conclusions from these. I know of the photosynthesis that it is a process, that produces energy-rich substances, such as oxygen from low-energy inorganic substances, such as carbon dioxide and water.

This process takes place in three steps:

Firstly, the electromagnetic energy (light or heat of suitable wavelength) is absorbed, using dye (chlorophyll).

Secondly, electromagnetic energy is converted into chemical energy. Light is used to enable electrons of water compounds to a more energetic state, and to split water. Electrons are transferred on chemical reactants, i.e. to the low-energy carbon dioxide and finally, energy-rich molecules (via chemical energy) are formed.

In the third and final step, chemical energy for the synthesis of organic compounds is used to serve living beings for both-growth and energy metabolism. In this phase, mitochondria come into play, synthesizing internal adenosine triphosphate

(ATP). ATP is the universal energy carrier in each cell and also an important regulator of energy-yielding processes.

Photosynthesis runs particularly efficiently, when - in the first process step, the absorption of light by an attunement of the pigment molecules (chlorophyll) is optimized with each other. In the second step, for the splitting of water molecules, energy of four stimulated pigment molecules is required.

It was found to be fact that stimulations of pigment molecules of sea-weed at room temperature, occur coherently in a way that can not be explained by interactions of classical physics. The observed high efficiency of photosynthesis obviously comes about through quantum entanglement.

Which conclusions can be drawn from this? An inductive conclusion from the particular to the general would certainly be wrong. To arrive at correct conclusions it is necessary to find out, does represent the most general processes in the cells of the sea-weed, due to biological-processing. Then, deductive conclusions to processes of the same species in other organisms are permitted.

In my opinion, the high efficiency of photosynthesis is not exclusively limited to marine algae. Photosynthesis generally occurs in plants, but also in bacteria (e.g. cyanobacteria). Moreover, as we mentioned further above, a marine gastropod, Elysia timida, was found that uses photosynthesis from incorporated plastids. If we find in all these cases the same type of photosynthesis, as in the case of the investigated sea-weed,we can conclude with a clear conscience that quantum mechanical entanglement provides for high efficiency in photosynthesis.

More difficult are conclusions on processes, that have nothing directly to do with photosynthesis. Here the complex photosynthesis must first be reduced to more simple processes, due to their properties. One of the characteristics of photosynthesis is their high efficiency, which can not be explained by classical physics. Further characteristics are the absorption of light initiated by certain molecules, and the possibility of **coherent excitation of molecules**, to initiate or optimize biological processes. If a lot of such properties are essential for the occurance of quantum mechanical entanglement during photosynthesis, properties are not that much special that they differentiate the sea-weed from other biological entities, but rather something general, indicating quantum entanglement.

If properties of the same type in other biological processes are found, conclusion on entanglement works again and is allowed without abdominal pain.

Can it be inferred from the known that - perhaps, quantum mechanical entanglement in genetic materials is responsible for a high efficiency of DNA replication?

Plants and algae have DNA, operated in their photosynthezising organelles. In mitochondrium, an organelle enclosed by a double membrane, there is another extrachromosomal DNA. Experimentally it is also known that DNA is somehow capable of absorbing light.

At the beginning of the replication process, the double-stranded DNA is first cleaved by the enzyme helicase. The binding forces between the individual strands must be overridden, however. This occurs only above a certain temperature, called the melting point. It lies close to expect then, that the absorbed light serves to achieve the melting point, whereas helicase controls the process of splitting, because chemical energy is no longer needed to achieve the melting point. And right in photosynthesizing cells electromagnetic energy is available and is even bundled and transported with the aid of quantum mechanical correlation.

For the purpose of replicating, each short DNA fragment is synthesized according to its model. When the replication is completed, the enzyme ligase links together new DNA-segments into one new strand. Overall, it is a complex production process, during which a lot can go wrong, unless sub-processes are tuned quantum-mechanically with each other, which means, they must be entangled.

Looking at my choice of words, we currently do not know with absolute certainty, whether quantum mechanical entanglement in DNA replication works at all. Hence, we are forced to guess. Assuming, DNA-replicase is accompanied - by classical evaluation not satisfyingly explained high efficiency of the process, and only coherent excitation of the molecules is involved in the processes to optimize the overall process, our guess is correct with a high probability.

The question, whether electromagnetic energy (=light) is utilized in mitochondria can be answered by the further above predicted. When light is utilized in DNA replication - and I assume that this is the case - then light is also utilized in mitochondria, because it contains DNA.

What can Elysia timida reveal about the evolution, and even of consciousness, I will explain later (see section 3.2).

@@@

Snails do not only steal chloroplasts,

but can purloin nematocysts from corals and sea anemones. The thumb-sized squid, Euprymna scolopes, incorporates a bacterium, Aliivibrio fischeri, that is able to glow. In his special luminous organ, the mollusc literally breeds bacteria and distributes them through fine channels in the skin over the entire body surface.

More at: http://www.tagesspiegel.de/wissen/diebische-tiere-bewaffnete-schnecken-und-leuchtende-tintenfische/10188290.html

@ KDS:

... how photosynthesis might work quanta-associated, you explained perspicuously. I am convinced that quantum mechanical entanglement will also be detected on the DNA level.

In Haeckel's sense, it is basically an evolutionary process. From this point of view, even evolutionary components for all attempts of an explanation must be put into consideration:

An octopus incorporates luminous bacteria and, later on, illuminates itself due to its needs. How, actually, do living beings manage to handle this problem, to live in complete darkness, the deep sea, for example?

@@@

Quantum coherence in photosynthesis for efficient solar energy conversion

Photosynthesis is the basic process, which connects the light of the sun with the life on Earth. New measurements show in detail, how coherent combination of electronic excitation and molecular vibrational states can lead to quick and efficient charge separation, which is important for further steps of photosynthesis.

Source:

E. Romero et al.: Quantum coherence in photosynthesis for efficient solar-energy conversion, Nat. Phys., online 13. Juli 2014; DOI: 10.1038/nphys3017

@ NW:

... I am now able to explain very well the high efficiency of photosynthesis in plants by a quantum-related point of view: that there is an interaction happening between a photon and a electron bound to an atom or molecule, the photon must have a greater discrete (or the same) amount of energy - as it is required by the electron, to raise its energy to a higher level. Such interactions are essential for the first two steps of photosynthesis. But - most of the arriving photons to the antenna molecules, respectively to the pigment molecules of the chlorophyll, have a lower amount of energy than needed. What happens to them? Do these photons only increase the kinetic energy of the molecular movement?

The response and ingenious solution of nature was always before my eyes, now I understand her. To activate the electrons for chemical reactions, the antenna/pigment molecules must absorb the energy of at least four photons. If these pigment molecules were not quantum-mechanically entangeled, the individually incoming photons of lower energy would only provide kinetic energy of the molecular motion, nothing else. Entanglement, however, means that the molecules behave in the entangled state as a unit. The amount of energy of the incoming photons can be added for as long as needed, until the power is sufficient enough to interact with an electron.

Therefore, the **entanglement of organic molecules** seems to be a universal principle of nature: wherever nature captures photons, this principle can be found with the aim to collect their energy and use this energy for chemical processes. Only by pooling through entanglement antenna, molecules achieve necessary discrete amounts of energy.

@@@

Quantum entanglement in photosynthetic light-harvesting complexes.

Light-harvesting components of photosynthetic organisms are complex, coupled, many-body quantum systems, in which electronic coherence has recently been shown to survive for relatively long timescales, despite the decoherencing effects of their environments.

See more at:
http://www.nature.com/nphys/journal/v6/n6/full/nphys1652.html

@ KDS:

... the entanglement principle of organic molecules to produce energy to control essential life processes, such as the promotion of growth in plants, is very plausible. At the same time, however, it becomes clear how Bose-Einstein-condensates[36] come about even at room temperature. What you recently developed as a universal principle, researchers now confirmed experimentally:

Thus, Fleming and his colleagues[37] investigated the light-harvesting complex of green sulfur bacterium Chlorobium tepidum. The bacterium takes photons with its light-harvesting chlorosom-antenna[38] and relegates the excitation energy, thus released, in an electronic form to a reaction center, where then the actual photosynthesis takes place. The energy transfer from antenna to reaction center extends through the so-called Fenna-Matthews-Olson- or FMO-complex, which consists of three identical proteins, each with seven bacteriochlorophyll-molecules or chromophores[39].

In its natural habitat, only few light falls on these bacteria, so they need to perform photosynthesis with a very low light intensity. Therefore, an FMO-complex always only needs to pass individual stimulations while jumping from one chromophore to the other. Since these molecules have different large distances from each other, they are also differently strongly coupled with each other.

Particularly strong is the coupling with proximately adjoining chromophores, which form a dimer[40].

36 **Bose-Einstein condensates** are macroscopic quantum objects, in which the location of individual bosons is not concentrated on one position. This results in the property of superconductivity.

37 see http://www.cchem.berkeley.edu/grfgrp/

38 As **chlorosomes** are described intracellular organelles of photosynthetic green sulfur bacteria.

39 **Chromophore** denotes this part of a dye which ensures color.

40 A **dimer** is a molecule or a molecule composite, which often consists

Researchers at Berkeley have recreated the events in FMO-complex as realistic as possible. In the model, the earlier measured excitation energies of the chromophores and their energetic couplings were entered. Accordingly, two adjacent chromophores (1 and 6), which formed a dimer, absorbed the incoming energy of the light-harvesting antennas and passed them on. The model also took into account that a stimulation from the chromophores could be transferred to the them carrying proteins, and thus went lost for photosynthesis.

At the beginning of the simulations, that were carried out at both, low temperatures (77 K), and high temperatures (300 K), chromophore 1 or 6 was stimulated. All other chromophores were not initially excited. The simulations showed that in less than 100 fs (femto-seconds) every seven chromophores were transferred into an entangled quantum state, so that none of them remained in a unique state anymore. The bacteriochlorophyll-molecules over a distance of 2.8 nm (nano-meters) away, were entangled quantum-mechanically. The excitation could be performed on each of the chromophores. Where it really happened, remained undetermined.

The folded state has been found to be relatively durable. At low temperature, entanglement lasted about 5 ps (penta-seconds), at high temperature at least 2 ps (pico-seconds). This time-span is long enough to prove predicted entanglement by femtosecond spectroscopy.

Whether it is also possible, though to measure the entangled non-equilibrium condition by state tomography in detail, as was done, for instance with folded equilibrium states of atoms, still remains open. The latter aspect is therefore of greatest interest to the question, how a condi-

of two identical sub-units.

tion- far from thermodynamic equilibrium, can be maintained as long as is necessary, in order to "live".

As we elaborated in: „*Leben aus Quantenstaub* - Life out of quantum dust"[41], there is clearly an evident universal property of photons: supplier of information and energy, and matter generator. Further above, you also described very well, how a chemical reaction works with the participation of the electrons (= matter generation) to let a plant grow. Parallely, an energy providing biochemical process is set in motion, by provision of ATP in mitochondria, thereby preserving life´s regulation, far away from thermodynamic equilibrium. Basic information via entanglement ensures the provision of energy borrowed from the metric-free vacuum, to set this process of life in motion at all.

In plants chloroplasts (plastid) obviously seem to have adopted the antenna function.

In nanotechnology, nanotubes[42] are used as spin- filters. With the help of these, electrons can be influenced in quantum-dots[43]. Seemingly, the function of plastids of the plant could even be "imitated" by nanotechnology.

41 Wrobel, Sedlacek (2014) *Leben aus Quantenstaub* – Life out of quantum dust, p. 31

42 A **nanotube** is an elongated hollow body, with a diameter of less than 100 nanometers. Carbon nanotubes have been thoroughly investigated.

43 A **quantum dot** is a nanoscopic material structure, usually made of semiconductor material. Charge carriers in a quantum dot are so far restricted in their mobility that their energy can only assume discrete values. Shape, size, or the number of electrons in quantum dots, can be influenced.

2.5 Relation between gravity, entanglement and life-processes

@ NW:

... as we consider in our dialogue the processes of life under quantum-physical aspects, we should deal again with gravity. Unfortunately, there is no conclusive quantum theory of gravity. But I found an interesting addition to our dialogue in a little book *"Leben aus Quantenstaub* - Life out of quantum dust, engl.". On one particular internet-site,[44] numerous links to pdf-texts could be found, that were mostly written by Karl Otto Greulich. Greulich proposed a theory, which seems very plausible to me.

He suggests that **all elementary particles are formed by electrons**.

His arguments are strong and, because of Occam's razor (one of the most important principles for the establishment of scientific theories!), I rather tend to discard the still very complicated standard theory of elementary particles, which is kept alive by interest-driven arguments of mainstream physics. Greulich´s theory fits well to what we are discussing: we had namely imaginated that elementary particles arise, or incur out of photons[45].

Because of their size and related energy demand, it rather seemed unlikely to us that elementary particles could be created directly, by fluctuation of the metric free vacuum.

But now, for the development process of elementary particles, an intermediate step -with consideration of electrons, can be introduced:

44 see http://www.fli-leibniz.de/www_kog/research/physics/physics.html

45 see Zeit-article fom 17.10.1997
http://www.slac.stanford.edu/exp/e144/diezeit.jpg

fluctuation / decoherence -> photons -> electron -> proton, neutron, etc. -> ...

In addition, Greulich found that each gravity can be expressed in form of electrical power. This means that:

gravity is probably nothing more than one form of electromagnetic power.

The hitherto unexplained gravity could suddenly be explained very simply. I think it is also necessary to supplement my booklet "Sedlacek: *Supervereinigung* - Super unification, engl.", because I will replace Greulich´s formula by the mass equivalent information.

The consequences are far-reaching. The relation, found by Greulich, could also explain what gives elementary particles their mass. The "god particle", as said for the Higgs boson, would be just as superfluous to explain the mass, as well as the particle accelerator at CERN. Unfortunately, mainstream physicists prefer to invent new particles: in the meantime, five different Higgs bosons have allegedly been detected. The theorists will certainly be busy for years to find a suitable fit for their theory.

For an outsider it is unfortunately not understandable, what sort of facts inaugurated physics specialists present to an interested public. So, the only option is to believe in the presented research results of elementary particle physics, as though the results were some sort of religion - or not. Fortunately, the elementary particle physics has virtually no effect on the quantum physics, and you can therefore put the results, that were presented on behalf of the Higgs boson by mainstream, and the incomplete elementary standard theory into a book-shelf.

@@@

Forces out of nothing

Among the most famous "forces out of nothing" there exists one force, which, in 1958, was experimentally confirmed for the first time , the Casimir effect[46]. This effect was predicted in 1948 by the Dutch physicist Hendrik Casimir. The Russian physicist Jewgeny Lifschitz expanded in the 1950s Casimir calculations to more general cases. He also showed that the Casimir force does not only work as attractive, but also as repulsive power, depending mainly on the properties of the material. This prediction was experimentally verified in 2009. Researchers hope to profit from it by having objects levitate smoothly.

More at:
http://www.heise.de/tp/artikel/42/42230/1.html

@ KDS:

... if - along Greulich's view, gravitation can be interpreted as an electromagnetic force, the question arises, how, who or what mediates the specific interaction between the metric-free vacuum and the 4-dim world? In principle, there comes a whole class of electromagnetic waves of different wavelengths into consideration, for example microwaves or X-rays.

During the Casmir experiment, two thin metal plates are placed together very tightly and contactlessly in an engineered vacuum environment. It can be observed and measured that the plates move towards each other by acting

46 The **Casimir effect** is a physical quantum effect, which causes a force to act on two parallel conductive plates under vacuum, compressing both of them.

forces. Where do these come from? Is it not obvious to see their origin in the metric-free vacuum? The forces may be caused by an electromagnetic gravitational effect. Should research on the Casimir experiment find out, at which wavelength the forces could be associated with a gravitational electromagnetic phenomenon, another assignment would be possible. There seems to be no idea, however what gravity really consists of[47].

That there could be gravitational waves, as advocated by Einstein, and later confirmed by indirect detection through discovery of a new pulsar type by the astronomer Joseph H. Taylor and Russell A. Hulse, Princeton University (New Jersey), seems to be rather a confirmation of the general theory of relativity. Thus, gravity would be a property of space-time geometry[48].

Closer to reality concerning our dialogue were the considerations of De Broglie (1929), who, analogous to the photon double-slit experiment, assigned quantum properties to electrons[49], which could be confirmed experimentally by a double-slit experiment by Jönnsen in 1959[50].

Overall, a picture emerges inside myself, according to which gravity could de facto be regarded as a kind of electromagnetic force, whose mediation is assigned to photons for obvious reasons that have arisen from the metric-free vacuum by fluctuation and decoherence.

Gravitation would therefore be generated by one of the universal properties of the photon, because - in our opinion,

47 see http://www.3sat.de/page/?source=/nano/glossar/grundkraefte.html

48 see http://www.spektrum.de/alias/dachzeile/nobelpreis-fuer-physik-indirekter-nachweis-von-gravitationswellen/821271

49 see http://de.m.wikipedia.org/wiki/Materiewelle

50 see http://de.m.wikipedia.org/wiki/Claus_Jönsson

photons form time and space, and gravity[51] can also be interpreted as a property of space-time geometry.

In an analogous cosmological context, gravity would be produced by photons and disseminated together with the background radiation in the microwave range.

Now, you mentioned previously:

> *"... Greulich found out that each form of gravity can be expressed in a form of electric power. This means that gravity is probably nothing more than a form of electromagnetic power. The hitherto unexplained gravity could suddenly be explained very simply".*

In Greulich´s formulas, only the mass would have to be replaced with equivalent information

With these considerations, the assembled “thing” gains round contours: fluctuation/decoherence certainly “act” as the smallest unit, namely the Planck constant of a photon with very different properties, such as

- **matter generator** via interaction with electrons: matter is established
- a form of electromagnetic forces: to **generate structure-gravitational forces**
- **power supplier**: animate beings are kept alive.

Crucial basic size could only be basic information, being stamped in the metric-free vacuum and stored there[52].

---

51 see Wrobel, Sedlacek (2014) *Leben aus Quantenstaub* – Life out of quantum dust, engl., p 43 ff

52 Wrobel, Sedlacek (2014) *Leben aus Quantenstaub* – Life out of quantum

@@@

@ NW:

… very interesting article! As you rightly say, the thing is getting round. Casimir forces in conjunction with Greulich´s formulas could continue to shed light on the darkness: maybe we really are on the track to the essence of gravity. Will continue to reflect on it...

@@@

Critical Casimir effect in colloidal model systems (doctoral thesis)

Both, the state and the physical properties of matter, are directly dependent on the manner, in which its components interact with each other. In the field of soft matter - colloidal suspensions also included, there are various interactions. Many of these interactions can to a certain degree be controlled.

The critical Casimir interaction is an interaction, that can be induced in critical binary mixtures, and depends ultra-sensitively on both, the temperature of the sample, as well as the nature of the interacting surfaces.

Colloidal systems can be regarded as structured anisotropic colloidal particles, the so-called Janus particles. The chemical structure of surfaces allows generation of directional interactions within the system, that can be induced in the vicinity of the critical point of the solvent. Due to anisotropy of interactions, formation of ordered structures of

dust, engl., p 116

high complexity is expected in such systems, during the self-assembly process[53].

More at:

http://d-nb.info/1022562460/34

@ KDS:

... if I understood you correctly, you want to design a comprehensive theory of the gravitational term with elementary information as follows:

> *"... calculate on the basis of the formulae, found in a general physics book for scientists and engineers, the gravitational force between two free, non-polar atoms / molecules. Subsequently: the resulting electrostatic force of attraction between the atoms / molecules on the condition of an induced dipole moment for both atoms / molecules.*
>
> *If the various forces (gravitational and electrostatic attractive force) under the stated conditions are the same, or at least of the same order, then a theory of the gravitational force would have been found ... ".*

A complete theory also includes to experimentally verify, or falsify it. To me, gravity seems to have something to do with electromagnetic forces. Where these forces come from, and how they could be standardized and made quantifiable - is not clear, yet.

53 **Self-organization** describes a system, which controls, orders, forms and restricts elements and processes, and which exists without recognizable external controlling elements.

During the Casimir experiment, thin metal plates can move together in a vacuum environment, comparable to attraction. It is permissible, to assign the quantum effects and the hence generated forces to a technically produced vacuum environment. This raises the question, whether the described effects occur only with metal plates? Apparently not, because in the meantime, these effects can be detected in all sorts of shapes and materials. In the doctoral-thesis described above, these effects are also true for colloid systems.

In experiments with particularly thin metal plates the suspicion arises that there could be a mass attraction in terms of a gravitational effect. If so, the necessary forces can only derive from the vacuum. Do you think, this is possible? Is the Casimir effect an experimental proof of your theory of gravitation from elementary information?

The presented doctorate work reports as an outlook, that Casimir effects could be made technically available even with carbon nanotubes. Immediately, a comparison with biological forms - such as humans, imposes.

As we know in the meantime, man can be interpreted as a **macroscopic quantum objec**t and as a structural space-time construction within the 4-dimensional world. Consciousness-driven interactions with the metric-free vacuum permanently occur. Although being actually "hollow" in a quantum physical sense, condensed matter presents itself as a stable structure. Where else, if not from the metric-free vacuum, do the forces come from, to ensure things like these?

@@@

@ NW:

… you found a very interesting thesis. But I am afraid, it cannot help us further in the context of gravity, and here is why: Casimir forces result in the immediate vicinity of a shield. In between the metal plates used in the Casimir experiment, not all wavelengths (via photons) can be existent; outside the plates is principally possible. The photons outside the plates produce therefore a greater pressure on the metal plates than the inside ones.

The further the metal plates are away from each other, the less counts the considered description. This would consequently mean that nearly all wavelengths also occur between the metal plates.

The same applies analogously to the mode of action of Casimir force in liquid mixtures, next to a critical point. The effect is closely limited to the range of the liquid.

Gravity, however, is a force acting over huge distances. If you want to describe gravity neither classically, nor with the general theory of relativity - which is indeed a classical theory, then only one way remains to conceive gravity as a kind of electrostatic force effect. Why?

Principally, on electrostatic force applies the same inverse square distance as is valid for gravity. Although it was always suspected, gravitational and electrostatic force to basically be the same - classical physics could not justify. On a quantum level, it would be possible, though. This is precisely what I have been working on: can a connection be made between the electrostatic forces from induced dipole moment and gravity?

Unfortunately not. My calculations did not confirm such a connection.

Electrostatic interactions occur through interaction of unmoved electric charges. Hence, I again made a rough calculation with the aim: find an action of force, similar to a gravitational force between the electrostatic field of the Earth, and the induced electrostatic field of a lead bullet. Unfortunately, the amount of both force- effects differs so much that equality is virtually eliminated, which means very clearly: gravitation cannot be an included electrostatic force in our known manner!

On the other hand, gravity can not be produced by general electromagnetic forces and moving charges. For, if there are no moving charges, gravity can not be the result of known electromagnetic forces.

Now to my new idea: all atoms originally contain elementary particles with spin (spin, angular momentum): electrons, protons, as well as neutrons, have spin, but photons have so, too. Important experiments on the spin behaviour, based on the spin property of charged particles, assigned to a magnetic moment.

In the case of the **Einstein-de Haas-effect**, the change of direction of the electron spin puts an iron rod into a macroscopic rotation. The **Stern-Gerlach-experiment** allows the first ever direct detection of direction-quantization through electron spin.

The effects of nuclear magnetic and electron spin resonance are used in chemistry, biology and medicine, for detailed investigation of materials, fabrics and processes. The magnetic moment m (also **magnetic dipole moment**) is a measure of the strength of a magnetic dipole in physics. A magnetic field belongs to each magnetic dipole. This means that even non-magnetic materials, or a body, must have a magnetic field, albeit a very weak one.

Consider now the Stern-Gerlach-experiment in detail:

A beam of (electrically neutral) silver atoms traverses the gap between the poles of a magnet in a vacuum environment. One pole piece has the form of a jet, parallel to the cutting edge, the other one of a flat channel; the magnetic field is thus strongly inhomogeneous transversely to the beam. On a glass plate, the silver precipitates. There are two distinct spots found, that is the magnetic field splits the beam into two separate sub-beams.

My comment: although silver is not a ferromagnetic material such as steel, it is influenced by a magnetic field.

If all bodies generate weak magnetic fields (magnetic dipole moments), then the effect we regard as gravity, could be an interaction of the magnetic dipole moments of matter, respectively of bodies. It would even be possible to assume very plausibly, the orientation of the dipole moments of two different bodies through the mediation of entangled photons. Bodies generally attract each other: photons have probably a spin with a magnetic moment.

With additional consideration of my theory, that space and time (space-time) are being generated by photons[54], and Einstein's brilliant performance to represent gravity as a property of space geometry, it is possible to devise a theory in combination with the above-said, that describes gravity as a sort of (electro-) magnetic force action, which is also generated by photons.

And - with the fact, that a photon can be viewed as condensed information from the metric free vacuum, we found an explanation for a possible connection between gravity and life processes, assuming that these are closely connected with

54 see Sedlacek (2010) *Supervereinigung* – Superconduction, engl., p 47

photons. The latter assumption is confirmed by experiments described in a later chapter (4.5). Speculatively, I can even go one step further, to assume gravitation being possibly mediated by entangled photons.

When I however consider that it took Albert Einstein more than ten of his best years to link gravitation mathematically with the space-time geometry, by the field equations of general relativity, I had rather younger ones to elaborate on a math-model of the new idea - thinking of my age!

@@@

@ KDS:

... in an internet video, a lot of things are shown, the quantum mechanical phenomenon of entanglement for example, and some of the things you already described. Interestingly, the Stern-Gerlach-experiment with magnetic dipoles compares properties with those of electrons or photons. I am convinced that the "magnetic feature" of the electromagnetic wave of a photon forms a force that gives all quantum objects a structure, namely the gravitational force.

2.6 Energetic cohesion by tunneling effects

@ KDS:

... we used the experiment by Nimtz[55] as an example for a successful tunnel effect[56]: Mozart's symphony was - although of a poor quality, nevertheless audible in the forbidden zone. Suppose, analogous to the cosmic background, radiation in the microwave range, exactly the same happens in the experiment of Nimtz, but this time in connection with the development of our 4-dimensional world. To proceed further, assume there are photons, that appear in the forbidden zone. Then you may conclude that only the photons in our 4-dimensional world appear to have been able to overcome the potential barrier. Do they subject now to mere chance to overcome the potential barrier by photons, or do there have to be special properties, such as the presence of a specific wavelength, or another peculiarity, for instance an occasion, or a requirement, in order to overcome the barrier?

As we assume, photons are considered to be the building

55 **Nimtz** is a German physicist (* 1936) who primarily has become known by his attempts to faster-than-light- experiments, using tunneling phenomena. In 1994, he undertook, together with Horst Aichmann a spectacular experiment, in which microwaves of the 40th symphony by Mozart were imprinted. This music signals on microwaves were transmitted through a barrier in the waveguide. It turned out that modulation of music on the microwave spread 4.7 times faster than light in a vacuum (faster than light).

56 see Wrobel, Sedlacek (2014) *Leben aus Quantenstaub* – Life out of quantum dust, engl., p 19

blocks of the original 4-dim-world[57]. Through information-processing a matter-containing-reality is created instantaneously.

In principle, to overcome a potential barrier separating reality from our metric-free vacuum, all possibilities come into consideration: mere chance, a specific wavelength, or an interaction. Probably, an interdependent process is likely to be set in motion, though: on the one hand, fluctuation and decoherence subject to an objective and not influenceable chance, namely mere chance. On the other hand, photons with a wavelength of 300 mm to 1 mm (microwaves) occur preferentially in our universe, which is evidenced by observable - and therefore measurable, cosmic background radiation.

There remains the question, whether there can be a cause, or a requirement, that allows specifically photons to overcome the potential barrier to create reality?

It is possible to overcome such a potential barrier using information-processing, which can be triggered in principle by all quantum objects of the 4-dim-world, especially, if it meets the criteria for elementary consciousness.

In this context, could tunneling effects and decoherence - as the universal principle, that condense and stick together matter – not be accepted with all organic and inorganic matter?

The fulfillment of the criteria for elementary consciousness[58] can be in principle confirmed on all dependent, on merely coincidenced quantum processes, including tunneling, namely on all animate beings and inanimate matter. As for Haeckel, the differences of his ideas to ours are not wide

57 see Wrobel, Sedlacek (2014) *Leben aus Quantenstaub,* Life out of quantum dust, engl. p10

58 see p 66

(see chapter 2.1.), especially, when I look at the following quotes:

> *... properties of the organic occur even in inorganic matter and are hence common to all natural bodies and, ultimately, the atoms*
>
> *... each atom has an inherent amount of force and in this sense being "inspired"*
>
> *... without the assumption of an "atom-soul", phenomena of chemistry or physics, such as attraction and repulsion, or movement of atoms, in the sense of feeling and will, are inexplicable.*

To sum up on the background of this context: a very fundamental question arises inside my self again: "What is life?" - the fundamental question arises and - most importantly, how was it created?

@@@

Quantum Tunneling to the Origin and Evolution of Life

Quantum tunneling is a phenomenon which becomes relevant at the nanoscale and below. It is a paradox from the classical point of view as it enables elementary particles and atoms to permeate an energetic barrier without the need for sufficient energy to overcome it.. This review shows that quantum tunneling has many highly important implications to the field of molecular and biological evolution, prebiotic chemistry and astrobiology.

Source:
Current Organic Chemistry, 2013, 17, 1758-1770.
Quantum Tunnelling to the Origin and Evolution of Life.
Frank Trixler, Center for NanoScience (CeNS), Ludwig-Maximilians-Universität München, Schellingstraße 4, 80799 München, Germany

@ NW:

… tunneling effect is an illustrative designation for an idea that one atomic particle (e.g., electron or proton) can even overcome a region of space (correctly: potential barrier of finite height), when its energy to do so is not sufficient. In fact, atomic particles have wave properties, too, and are more or less everywhere at the same time[59], which is described in terms of quantum-physics by their physical probability. Through this, they can be found in areas where their energy is actually not sufficient. But electrons, or protons and other particles, are able to borrow energy from the metric-free vacuum.

The borrowed energy must, however, be paid back after a short time. The tunnel effect means, particles can be found outside of the atoms or molecules. When protons or electrons are outside the space region of an atom (i.e., the potential barrier has been overcome) some time, they do not need the borrowed energy any longer.

In 1926/1927 the effect on isomeric molecules was first discovered and described by Friedrich Hund. In biology, it was demonstrated these days that the occurrence of proton tunneling in DNA is responsible for spontaneous mutations[60].

59 see http://idw-online.de/de/news593979 - Kassel physicists demonstrate: electron occurs simultaneously at two different places.

60 Per-Olov Löwdin: Proton tunneling in DNA and its biological

Proton tunneling is a general quantum physical effect (the same applies for electrons), which resulted into many technical applications. The same effect occurs not only at specific DNA structures, but also everywhere in biology, where similar structures, including plasmons (= extrachromosomal proportion of the genome) are found. Quantum mechanics tell us in addition that proton tunneling occurs everywhere, in all atoms, although in most of them less frequent.

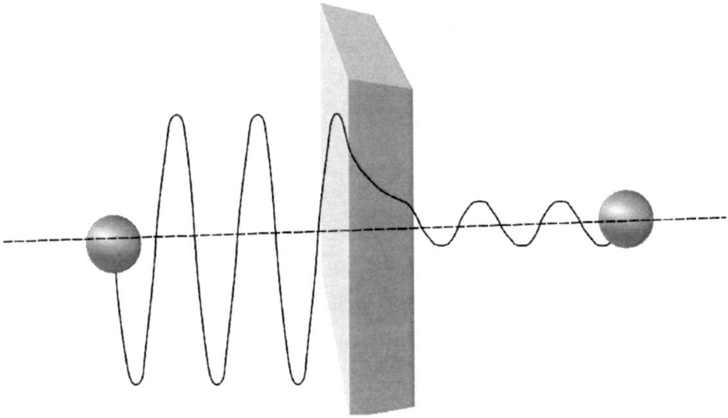

Fig. 4: Quantum tunnelling or tunneling (see spelling differences) refers to the quantum mechanical phenomenon where a particle tunnels through a barrier that it classically could not surmount. CC0 Yoschi

implications. In: Reviews of Modern Physics. . 35, No. 3, 1963, pp 724-732, doi: 10.1103 / RevModPhys.35.724

3. Animate vs inanimate: differences/implications

3.1 Consciousness as a pre-requisite for the evolution of life

What is life?

Erwin Schrödinger: "... The similarity [between clockwork and organism] is based, quite simply, on the fact that the organism is also anchored in a solid body - the aperiodic crystal, that forms the genetic material and is largely removed from the disorder of thermal motion.

... The most characteristic essentials [of the difference] are: first, the strange distribution of the "gear wheels" in a multicellular organism, and secondly, the fact that the single gear wheel is not a clumsy man-made, but the finest masterpiece that ever, according to the principles of God's quantum mechanics, has been completed."

See more at: http://online-media.uni-marburg.de/biologie/genetik/boelker/Wissenschaft/Erwin%20Schroedinger_Was%20ist%20Leben.pdf

@ KDS:

... Schrödinger was a pioneer in DNA-coding. He inspired Watson and Crick, who later found the DNA double helix and won the Nobel Prize for their discovery.

As is known, a large part of the chromosomal DNA represents **statistical information**. This type of information - due to our speech and way of thinking (syntax, semantic), is not equivalent to matter or energy, and therefore can not exist by itself, but requires a material or energetic carrier[61]. We assume that our 4-dim-world is being formed by the metric-free vacuum. Included into this statement are both: organic and inorganic substances. In a quantum-world, these substances are regarded as quantum objects through specific interactions, such as tunneling effects, or by entanglement. They draw energy from the metric-free vacuum, or borrow it and by doing so, generate a matter-containing reality. This is implemented by information-processing, which presupposes the existence of an elementary kind of consciousness[62] in each quantum object.

How to imagine the development of a chromosomal DNA to a finished life?

Archaea, for example, live deep in the Earth. They have never a chance to relate light energy to sustain life or reproduction. Something similar happens to deep sea creatures as well as **cave fish**, which have all found ways to live in an otherwise hostile environment. They all must equally have access to something that allows them to exist there.

Finally, the common denominator of all three listed species aims at a "target-orientated behaviour", here with the very clear will to survive/live in an otherwise hostile environment. They must apply somehow emotions[63] that lead to their target-orientated behaviour. This behaviour does not necessarily have to be objectively "reasonable", but can

61 Wrobel, Sedlacek (2014) *Leben aus Quantenstaub*, p 61

62 For the term 'consciousness' see box on page 66

63 "...**emotions** are states in which - even if only temporarily - all physical resources for a particular purpose are activated, and in which the attention [...] is directed very specifically to this goal"- Dawkins (1994), p.212

also be simply inked subjective. For this behaviour, physical resources are provided that are activated for a specific purpose, in the sense of: non-deterministic choices between alternative courses of action. "Emotions", however, imply a subjective or phenomenal view of consciousness. Arguing objectively, it makes sense to use the term "**need**" an "inclination to pursue a specific goal": as a whole, it is a **"goal-directed behaviour for the satisfaction of needs," to want to live** in an initially hostile environment, as is shown in the chosen example.

If we consider consciousness as an entity from an intentional point of view, and want to explain it the same way, too, then the specific „behaviour" must be part of the consciousness concept: it is "the (conceptual) anticipation of action steps to achieve a goal."

In terms of planning: in the case of new requirements, or changed external circumstances, undetermined decisions must be made between alternative courses of action. Had the decisions to equivalent alternatives clearly been predictable, which means deterministic, the opposite of consciousness would be fact, namely something automated[64] .

Now, we can assume of the above three species that they must have some sort of consciousness to live at all, and - to top it, live in hostile environments. The energy they draw for their life and reproduction, they trigger through information-processing, involving the metric-free vacuum, which then provides energy to survive. But consciousness does not seem to be reserved exclusively for animal life, but also refers to the quantum level both, organic as well as inorganic substance.

The "Atomistische Theorie des Bewusstseins - Atomistic theory of consciousness, engl." or the "Zell-Psychologie -

64 Sedlacek (2012) *Der Widerhall des Urknalls* –Echo of Big Bang, engl., p 154 ff

Cell psychology, engl." (Haeckel, Die Welträtsel), or the behaviour of photons during the double-slit experiment, supports quite explicitly the idea of the existence of a set of quantum objects to be assigned to "quantum consciousness".

Max Planck expressed the same view[65]:

> *„It is a conscious mind, which applies the power to move atomic particles to vibrate and hold the matter."*

Consciousness is something elementary. It brings forth matter, space, time and life.

What we commonly call evolution, occurs later on. It can be regarded as a slow, continuous, progressive development. In biology, evolution is the phylogenetic development of organisms, starting with the first occurring creatures to the highly developed species, such as human beings. In contrast, the cosmological evolution is baptised so after the Big Bang, as for the expansion and development of the universe (in the classical physical sense). With increasing knowledge, such as: elementary processes generate effects, it is allowed to speak of an **evolution of consciousness**[66][67].

Quite surprising is the simplicity of the recognizable control mechanisms in biological evolution: in genes encoded inherited characteristics (= statistical information) of a population change through the evolutionary process from generation to generation. This can be attributed to just three principles, that are iteratively gone through over and again:

In an undirected random process (= mutation) different

65 Planck (1989): *Physik und Transzendenz*

66 Wrobel , Sedlacek (2014) *Leben aus Quantenstaub*, p. 89

67 Görnitz, Görnitz (2009) *Die Evolution des Geistigen*

variants are produced from something that exists, to something completely new.

The second principle is the coupling of features of two successful individuals of one population (= recombination) to something new, or randomised better features.

The selection finally evaluates the results of recombination and selects, or is preferably based on, current environmental conditions. The process gets a target direction insofar, as those individuals, whose features are less advantageous with respect to the environmental conditions, have lower chances to survive[68].

Due to our speech and way of thinking (syntax, semantic), selection arises through information-processing of quantum objects involved. In summary,

- Elementary information
- Fluctuation (= mere chance) and decoherence
- Consciousness
- Evolution

represent the ingredients for the creation of all living.

On the basis of the unity of nature, a general validity of the principle described above is adopted. The quantum world as we know it, has therefore developed evolutionarily.

By information-processing, something abstract is linked to the instant physical reality and creates every single moment matter-containing reality.

In the 4-dimensional world, in which we live in terms of a space-time-construction (= body), our imagination suffices rudimentarily to understand "life". It is significantly more diffi-

68 Sedlacek (2012) *Widerhall des Urknalls*, p. 121ff

cult, however, to imagine cosmological evolution and information-processing of all the inanimate inorganic matter.

@@@

Terms

- **Consciousness** is ...
  ... **information-processing,** ...
  ...in the course of which **non-deterministic choices** between alternative courses of action are made and - in the case of new requirements, or changed external circumstances,...
  ... lead to **goal-directed behaviour** for the **satisfaction of needs**.
- A **need** is the inclination to pursue a goal.
- **Self-consciousness** is a higher form of consciousness, which is found in humans as well as in bonobos, e.g..
- **Primary consciousness** in living beings is information-processing below the level of self-consciousness, in which the criteria for consciousness are met. Among other things, subconsciousness is the primary form of consciousness.
- **Elementary consciousnesss** is an elementary form of consciousness which, meets the criteria for consciousness, and can be found in quanta, elementary particles, or in inanimate matter.

@@@

@ NW:

... I assume, as you do, that information-processing we identified as consciousness, is significantly responsible for survival in an otherwise hostile atmosphere, in an ice cave in deep-sea vents, gusher-sources, or other extreme habitats. Without the aid of a consciousness-process, it would certainly not be possible to adapt to external conditions, that are drastically different from habitat to habitat, definitively not always constant, but are subject to fluctuations.

It seems to be fact that the community members in their extreme habitats must make their own undetermined decisions in the terms of their own needs, to stay alive and to replicate.

After the criteria for consciousnesss are apparently met, we can not help but to say that consciousness-processes are even essential for the survival of the bottom life of our earth.

We can discover neither any kind of brain, nor place in the archaea, or other micro-organisms, though, that are particularly suitable for the processing of information. Like the double-slit experiment with photons, we are confronted with the question: where exactly does information-processing take place, then? Where exactly is the information store to find, et cetera.

As far as the purely physical processes such as replication are concerned, the necessary information for that is possibly stored in the DNA. But, detection of environmental changes requires the storage of a prior environmental condition. Without information of a former state, changes can not be recognized. And, if a living being does not recognize changes in its environment, it can not respond to it, it can not take decisions, it can not adapt. Without adaptation, it ultimately perishes.

Micro-creatures obviously managed to maintain and to adapt to most extreme habitats over extremely long periods of time. So, where is the updated environmental information stored? Where does associated information-processing take place?

At a different place, we already had to respond to the same question, and we had come to the conclusion that the metric-free vacuum, that is the nothingness, was the place where something was stored - where at least a portion of information-processing occured[69]. Similarly, we found that quantum physical processes, including entanglement, generally play a role in cells and DNA. Hence, this can not be different in extreme locations with micro-organisms and archaea.

Archaea, which are at the beginning of the food chain of a community, probably collect the energy they need to live, with the help of quantum mechanical entanglement of photon processes, that also occur in dark and extreme habitats. This happens probably - similar to photosynthesis, in the marine alga, which also makes use of quantum-mechanical entanglement-processes. The entanglement processes themselves are required to collect low-energy photons, until the amount of energy is sufficient, to put biochemical processes in motion.

The consciousness processes required for this, must probably be held to a considerable extent in the metric-free vacuum. Without a constant connection to the nothingness, as we also called the metric-free vacuum, nothing works for micro-creatures at all: no food, no replication, no adaptation to changing environmental conditions, no life-preserver.

69 see Wrobel, Sedlacek (2014) *Leben aus Quantenstaub* – Live out of quantum dust, p 116

3.2 Does the sea-slug Elysia timida show consciousness?

@ NW:

… I would like to pick up again my already announced thoughts concerning consciousness. I had them when I interested myself for Elysia (see chapter 2.4.):

What does Elysia timida tell us about evolution and consciousness?

Sea-slug Elysia timida found a way to make photosynthesis of algae available. How is that possible? As natural-scientists, we assume that all complex biological systems were formed by evolutionary processes.

An evolutionary process consists of three steps, which I want to characterize again briefly: first, something new is created, possibly unprecedented. In the second step, the new is combined with the existing and presented for selection. In the third and final step, a selection is made from the offered. The selection can be done passively by interactions with the environment, or actively by consideration of an individual inclination to pursue certain objectives (= needs).

Sacoglossa, to which Elysia timida belongs, feeds almost exclusively on algae. The digestive organ of the slug crushes and breaks the eaten algae. A possibly new feature (see step 1 of evolution above) is that the slug, especially its bowel, is able to differentiate between different cell components of the decomposed algae. Via which biochemical functions or receptors this is made, is outside my own subject area; I do not speculate. The slug obviously has the possibility to selectively digest certain cell components, or maybe not, even though the individual components do not differ in principle, and other

Fig. 5: A typically sacoglossa , herer: "Oxynoe olivacea"; Foto CC-BY-SA 2.5 Regiomontanus

marine gastropods, despite the different algae components, digest the complete alga.

For the second step of the evolutionary process, this results in the following combinations: a. digest all cell components, b. digest chloroplasts, c. digest everything - except chloroplasts.

In the third step of the evolutionary process, there is a choice among the three options presented. With a **passive selection** by the environment, either everything remains the same (combination of a), or the new one is usually a crucial advantage for sustaining life and reproduction. By an **active selection**, needs can determine the choice, and b or c can come to fruition.

Biologists act on the assumption that, in periods of starvation, the Elysia-slugs are supplied with energy sourced from the chloroplasts, continue to operate in the intestine photosynthesis. An experiment showed, however, that the slugs are able to survive without photosynthesis in chloroplasts: after two months in darkness, the slugs were alive as before. Researchers now suggest, that the slug does not necessarily benefit instantly of the chloroplasts, but only at times, when the intestinal cells degrade them, e.g. in periods of starvation.

For a passive selection by the environment in the third step of evolution speaks the fact, that one can derive a marginal benefit to the life-preserver by the presence of chloroplasts in the intestine. But is this advantage a crucial one?

Against the action of a passive process speaks the ability of the slug itself, to recognize the difference of various cellular components of the algae, or through its intestinal cells. So there is something that can adapt to different requirements.

With regard to the selection in the evolutionary process, the selection of evolution fell on combination c, which means that everything is digested - except the chloroplasts. The slug seems to have the opportunity, however, to opt for the combination b in periods of starvation, namely, to digest the chloroplasts. There is a non-deterministic choice between possible alternatives of behaviour.

If we also assume that the slug has the simple desire to make its life more comfortable, by letting chloroplasts produce oxygen and sugar, then all the criteria for information-processing, I described in my work as consciousness, are met. The same process of consciousness that decides, when chloroplasts have to produce oxygen and sugar, and when both of the components have to be digested, conducted the active selection at the third stage of evolution, as well.

It is certainly not a sophisticated process of consciousness, as is the self- or upper consciousness in humans. It is rather a process similar to the subconscious. In humans, subconscious mind performs many choices and physical controls. Only the most important thing is fed to the upper consciousness for decision. And, what is most important, the subconscious will decide, anyway.

The Elysia-slug demonstrates us with a high probability that simple processes of consciousness themselves act on a not too high a level of development.

It is even likely to assume that one simple type of consciousness-process was involved in the ascent of life, right from the very beginning of the evolution. For pure random processes and without active selection it is more unlikely that the emergence of protein molecules, DNA, archaea, eukaryotes, mitochondria, or plastids happened.

By the way: the hidden things behind the keyword "auto-catalysis", hopefully lead to my better understanding of the origin of life. Catalytic formation of molecules is probably the beginning of emerging life.

@@@

3.3 De origine vitae

The origin of life

Ideas of science about the origin of life throughout history: spontaneous generation - hen-egg problem - who created who? - what is life? - primeval soup - provisional solutions ...

More at:
http://www.biobytes.de/sommerkurs/Lebensentstehung.pdf

@ KDS:

... Stuart Kauffmann[70] pointed out that in a sufficiently complex environment, i.e. in a region with a sufficient number of different molecules, an out-forming of auto-catalytic cycle processes becomes inevitable. In such a cycle, components of the process catalyze the formation of other components, which in turn catalyze further ones, until finally the formation of the first-one-mentioned becomes catalytically supported.

Auto-catalytic processes can form themselves spontaneously in a suitable environment, if only the energetic and physical drive is sufficient. This is to ensure that the auto-forming organization retains itself, and that the resulting entropy can be discharged into the environment, which is only possible in an open system. Random variables, or noise as an out-forming of an objective chance, indicate a quantum-correlated process during evolution, and allows us to assume that life is primarily based on quantum processes.

In contrast, Manfred Eigen favours auto-catalytic supply

70 Kauffman: *The oil droplets in water;*
http://www.spiegel.de/spiegel/print/d- 68525308.html

coupled to the processes (hyper-cycle)[71], from which a hierarchical order can be found. Easier cycles can be incorporated into broader, larger cycles. Finally, from developing uni-cellulars, organelles with a specific remit, such as mitochondria can emerge, which act as power plants. In ribosomes, proteins are synthesized, which act as components and catalysts. DNA, however, operates as data storage[72].

Without any doubt, the origin of life is regarded as a highly complex process that requires a number of ingredients to provide a robust form of life. It is, as already mentioned, to assume that quantum processes must necessarily have an effect, such as quantum entanglement in photosynthesis, which contributes to the life of plants.

@@@

@ NW:

... I would like to try to give an answer to your question, how quantum effects in plant eukaryotes are realized: basically, in biology, a complex, multi-layered system exists. In physics, especially in quantum physics, you have to do with relatively simple systems, for which usually mathematical model concepts exist, to predict their behaviour.

In microbiology, however, there is a lack of mathematical models with the consequence that the behaviour of microbiological systems must be derived from past observations and experiments. But I think that not of all components of these systems is known, what precisely they do.

71 Manfred Eigen, Peter Schuster: The Hyper Cycle - A Principle of Natural Self-Organization. Springer, Berlin, 1979

72 Görnitz / Görnitz: Kreativer Kosmos - Creative Cosmos

Even the smallest microbiological systems are still huge, compared with the quantum systems of physical experiments. In order to find a connection between these two areas (quantum physics and microbiology), we first need to have the knowledge, which components the examined microbiological system consist of.

Then we must find out as precise as possible, which functions, or tasks (processes) satisfy individual components. And only as the last resort, we can bring quantum physical objects and processes in correlation with the biological components and functions.

Mainly, it will be photons, but also protons, quantum entanglement and consciousness processes, that interact with microbiological components. Add to this the process of evolution, while life emerges.

You also yielded another piece of the puzzle by having made a connection to **auto-catalysis** in connection with the beginning of life.

@@@

Molecular Basis of Evolution (Schuster):

"Can we estimate the probability of the origin of life?...

...the natural history of the species and the hence resulting phylogenetic trees were fully confirmed by comparison of genetic information carrier and nowadays living organisms."

More at:
http://www.tbi.univie.ac.at/~pks/Presentation/duesseldorf-07.pdf

@ KDS:

... in the matter of auto-catalysis there is consensus on their existence as a prerequisite for becoming "alive". Eigen's hyper-cycle assumes that a single catalysis is not sufficient enough to cause a chain formation. Each catalysis always provides energy for the next. The system therefore perpetuates itself somehow. Only when a critical size is reached, "living" chain molecules are created[73].

For an auto-catalytic molecule, the minimal amount of information estimates the emergence of "living chain molecules", i.e., to the minimum complexity of the design. This is what Manfred Eigen and Peter Schuster have done. Their studies are based on experiments with the simplest self-reproducing specific molecules called phages (viruses that infect bacteria), and, additionally, on the mathematical interpretation of the experimental results.

The calculated size of this minimum amount of information leads to a dilemma. It is too big, as it could adequately be reproduced by a single auto-catalytic system of the simple type, as was possible in the early phase of the origin of life, with sufficient precision.

„Child's play“ is intended to illustrate the dilemma closer. The "Chinese Whispers" is suitable here to illustrate:

> *Some children are sitting in a row next to each other. The child at the beginning of the series comes up with a sentence created by itself, and whispers it to the kid sitting next to it. This child carries the sentence in a whisper to the next child on, and so the sentence runs through the whole series. Eagerly, the players want to know what happened with the initial sentence, when it has gone through the "Chinese whispers".*

73 http://www.chemie.de/lexikon/Chemische_Evolution.html

Was the sentence long, or complicated, it thus contained a lot of information, it will normally be unrecognizable at the end. The information was lost. A short, or not very complicated sentence, however - with only a little information content, will arrive without much distortion. It can even sometimes happen that the last child reproduces it more comprehensive than the original.

Eigen now determined for the original molecules of life an amount of information, that is needed to reproduce a living system, sustaining enough meaning to keep the entire machinery self-reproductive. And somehow, nature seems to have managed to overcome the dilemma between the minimum amount of information necessary, and the maximum possible error-free reproduction.

The trick was the **hyper-cycle**: several self-reproductive units, each of which had only to translate a small amount of information, which was below the maximum possible, co-operated with each other just like in a team, to reproduce the information required for the self-reproduction. And this co-operation was cyclical.

What does this mean? Our example of the "Chinese Whispers" must be extended to a „Super post" to get fit for the discussion. The first child whispers its sentence not only to one, but to several neighbours, as for the first generation. Each of these children whisper in turn their message to several neighbours, in the sense of the second generation, and so on.

The case is even more complicated: if a child received the sentence in a particularly understandable form, then the child passes it on to a lot of neighbours. But, if it receives a less useful version, the child loses soon pleasure when passing it on, so only a few children get the sentence. Completely misrepresentative sentences are no longer passed.

And thus the number of children do not grow into infinity, a master of any already informed generation takes a certain percentage of children from the game. Thus, the original sentence can optimize in a Darwinian way.

Finally, almost only very exquisite versions of the theorem remain. The less good ones are "extinct". Provided is here as before, that the original sentence does not contain more information than the one, that corresponds to the above-mentioned maximum amount. Otherwise, it was hopelessly lost after a few generations[74].

Where does the system now take the energy from for a hyper-cycle, for the production of chain molecules?

There is broad consensus that exposure to energy and, with the aid of catalysts, complex organic molecules were formed, which interacted with each other. Assumingly, there was co-operation, or **symbiosis** already in the prebiotic molecules. Eventually, processes similar to metabolism started to work in these structures, and they could multiply. They hence showed characteristic features of animate beings. It was probably a membrane that wrapped around these already highly complex proto-creatures.

Different ideas exist of the origin of energy for the growth of the proto-beings. In **heterotrophic lifestyle** they had won power by receiving energy from organic molecules out of their environment (De Duve: origin of life).

Autotrophic organisms get their energy from either the light by photosynthetic processes, or redox reactions[75] during a chemosynthesis.

---

74 http://www.zeit.de/1978/07/so-enstand-das-leben/seite-7

75 A **redox reaction** is a chemical reaction in which a reactant electron transfers to the other. Combustion is an oxidation-reduction reaction using oxygen as the oxidizing agent.

If the first living beings were autotrophic, it might have been only one type of chemosynthesis (Wächtershäuser: Theory of surface metabolism). For photosynthesis, proto-creatures lacked of the pigments, or highly complex molecules, to absorb light.

Several agreements exist on the order of formation of biotic molecules, such as proteins and RNA[76]. Similarly, there are several hypotheses about the nature of the catalytic process and whether, and in what manner, hyper-cycles exist in the replication of RNA. Various statements persist about the extent of random, playing a role in all of the processes, or whether the lengthy process of creation of life and the further development (evolution) of a determinism (teleonomy), or purposefulness is inherent[77].

Regardless of these considerations, structures are compared (e.g. amino acid sequences) with each other, with homologous proteins from different organisms, in order to better understand the degree of relatedness of the organisms. The idea was that, with the evolution of organisms, an evolution of composing proteins would be associated.

Nowadays, proteins are also subject to selection pressure. Only the rate of evolution of the proteins differs markedly from the speed of evolution of the organisms, from which they were isolated. This is mainly due to the fact, that in the environment of organisms other selection criteria are considered in the environment of the proteins. Their environment is the cell contents, and its composition is fairly constant.

The evolution of proteins depends primarily on the characteristics of the genome of the cell. Therefore, only those cells are able to survive whose proteins insert in the net-

76 **Ribonucleic acid**. Structurally, RNA of the DNA is similar. It is needed in biological cells for protein synthesis.

77 ftp://ftp.rz.uni-kiel.de/pub/ipn/SystemErde/08_Begleittext_oL.pdf

work of metabolism[78].

For several years it has also been known that many, if not most of the genes of eukaryotes and archaea are made from pieces, that are separated from each other in the genome by non-coding sections. This organizational scheme of the genetic material obviously withstood the selection pressure, because the aim was, to give rise to new proteins as quick as possible and, above all, to minimize loss.

What is missing, in summary, is a clear idea about the energy balance of the hyper-cycle. If we may assume that it is a closed system, we are balanced. The hyper-cycle resembles however very much a perpetual-mobile-motion.

The possible resolution of this problem[79] finds a surplus of energy in the real 4-dimensional world, that always arises, when in the metric free vacuum a "lack of information" is coined. Only the deletion of information costs energy[80]. Since the information store of the metric-free vacuum is never deleted, though, the excess energy remains within the 4-dim-world and thus a matter-containing reality shapes there.

Taking into account the resolution of the energy-balance-problem, and taking into account the background on top of our information and selection term, there is manifest evidence for the assumption that a hyper-cycle is a quantum-associated process, due to which life is made possible at all.

---

78 http://www.biologie.uni-hamburg.de/b-online/d42/42d.htm)

79 Wrobel, Sedlacek (2014) *Quantenstaub*, p 116

80 Sedlacek (2012) *Der Widerhall des Urknalls* – Reverberation, thought experiment "Maxwell's demon", p 100 ff

3.4 Evolution as a conscious self-organizing process

Self-organization (synergetics)

The theory of self-organization stands within the tradition of the most common known theory of physics about objects in time - the quantum theory. The debate about the consequences of the quantum theory for apperceptive options of reality, knowledge and consciousness, will go on for a long time.

See more at: http://www.diss.fu-berlin.de/diss/servlets/MCRFileNodeServlet/FUDISS_derivate_000000002158/05_kap4.pdf ?hosts

@ KDS:

... mutation can create something that has never been there before. This creative process seems to be subject to an objective chance. The recombination represents rather a more directed, non-objective random. By contrast, at the third step of evolution - the selection, a process is taking place which seems to meet the criteria for a primary or elementary form of consciousness[81]. Situatively, options of the "offered" possibilities are selected, and the best one is picked to meet with a particular situation (=selection), to be able to survive in extreme conditions (= target) - e.g. archaea show us that this strategy works.

The hypothesis that selection meets the criteria of consciousness can - by the way, be falsified: life is - under extreme conditions, also possible without selection, that is without a conscious selection process.

81 For the definition of consciousness see box on page 66

The theory of self-organization stands, by the way, in the tradition of quantum theory. So both, Carl Friedrich von Weizsäcker and Thomas Görnitz assume that the quantum theory is a general theory of the lawful behaviour of objects of every-day-experience (Weizsäcker 1991, Görnitz 1992). Not one single every-day-experience was found that would contradict the quantum theory in the context of the phenomenon of self-organization.

The postulated unlimited validity of the quantum theory for the phenomenon of self-organization, was confirmed by Herman Haken through the development of the laser light. Manfred Eigen confirmed it through the self-organized origins of genetic information.

Self-organization in the sense of synergetics (Haken 1977) refers to the ability of a system, to be able to make transitions between different structures, because of changes in environmental parameters, while no external instance is necessary for the structural re-formation. It is conveyed by the interior dynamics of the system.

The description of Haken comes very close to my own conception of evolution. In relation to the question: "What is life?", auto-catalysis can form chains and, by doing so, can produce DNA. It is a self-organizing process that is - in the frame of the definition, consciously **controlled by a kind of selection**. Inclusive of elementary information, the development of life can be interpreted as a triggered process due to information-processing.

3.5 Studies and reflections on inorganic life

Ernst Haeckel: Kristallseelen - Crystal souls.

Studies on inorganic life. Kristallotik (= Kristallkunde - crystallography), Probiontik (= Zytodenkunde), Radionik (= Strahlingskunde), Psychosomatik (= Fühlungskunde).

More at: http://caliban.mpipz.mpg.de/haeckel/kristallseelen/haeckel_kristallseelen.pdf

@ KDS:

...Haeckel writes, consciousness is neither more, nor less than any other mental activity, a natural phenomenon, and thus subjected to the law of substance like other natural phenomena. To apply the law of development of evolutionary theory onto inorganic matter, he had to furnish proof that inorganic substances, in principle, do not distinguish from organic matter. On the background of these postulations, he published in 1917 - in continuation of his lecture: "Zellseelen und Seelenzellen - Cell souls and soul cells" (1878) the font "Kristallseelen - Crystal Souls" (1917), already held 1878 in the Concordia to Vienna, to justify his theory [82].

In the chapter psychosomatics (= Fühlungskunde), he tries to prove that in organic and inorganic nature - everywhere, the same "eternal, iron and large laws" dominate all events. The whole "living world", inextricably linked to the so-called "lifeless world", is passed uniformly as a "universe", defined by the same principle of development. Later, he came to his conviction that the two main characteristics of the psyche: movement and sensation, are rather to be

82 http://digital.ub.uni-duesseldorf.de/vester/content/pageview/1966872

separated into two independent attributes of substance: into mechanics - with the movement, and psychosomatics, which deal with the perception of matter:

> *"All matter has life, inorganic as well as organic; all things are animated, crystals as well as organisms. Unshakably rises anew, the old belief of the inner uniform context, of all events of the unlimited power of universal laws of nature ...*
>
> ***... matter can never be effective without mind, never exist the mind without matter"***

To the naturalist, who conceives all phenomena as “nature”, the psychology of the ‚Seelenkunde' (psychomatics) appears as a unit of natural laws throughout the cosmos[83].

To maintain records the soulfulness of inorganic substances, Haeckel tried to show that crystals demonstrate properties of life, such as growth, nutrition, reproduction, skin formation, metabolism, nucleation, excretion, regeneration, movement and touch.

In our speech and way of thinking (syntax, semantics), and in other words, this means: all quantum objects of our 4-dimensional world have the ability to trigger information-processing as an expression of existing consciousness.

With this, and involving the theory of decoherence, it would be collectively and sufficiently well explained, why matter-containing reality also existed, if consciousness was disconnected from all organic.

@@@

83 see http://digital.ub.uni-duesseldorf.de/vester/content/pageview/1913746

Fig. 6: Crystal Souls - "Studies among the Snow Crystals ... " by Wilson Bentley, CC0

@NW:

… I am also enthusiastic about Haeckel`s mastermind-findings and can only agree to their contents. I slowly get the feeling not to be so very wrong with my own world-view. Via Pascual Jordan's "The Physics and the mystery of organic life" (Braunschweig, 1945), I try to get further confirmation. The quantum physicist Jordan represents at least the opinion that the physical laws of quantum physics are also valid for the processes of life.

@@@

@ KDS:

... in your book "*Der Widerhall des Urknalls* - The Echo of the Big Bang, engl." you deal with the topos of "consciousness units as elementary building-blocks of matter". It deals with the relationship between the phenomena of quantum physics, and information-processing. The latter operate as a link between physical reality and the metric-free vacuum.

We found that, with the help of quantum correlation (entanglement), energy is provided to a quantum world within the structure of a 4-dimensional one. All this is triggered by information-processing. The exchange can be carried out over a whole class of electromagnetic waves[84], such as microwaves, heat, or X-rays.

The smallest unit, that deploys an effect in our 4-dimensional world and gets real, is the Planck constant.

For obvious reasons, the smallest "active entity" is expected to be represented by a photon. Hence, photons are a vehicle for a discernible reality by sensory perception. The main sensation, with which animate beings like humans can make reality comprehensible, is carried out by quantum associated processes, e.g. by visions. Quantum biology will provide further evidence in this field in the years to come[85].

It can also be assumed that the 4-dim world immanently occupies a hybrid state: it is always the perceptible and ob-

84 Wrobel, Sedlacek (2014) *Quantenstaub* - Quantum dust, p.76)
85 see http://www.spektrum.de/news/mit-allen-quantenmitteln/1034267

servable in reality and, at the same time, non-local, connected by quantum processes in the metric-free vacuum[86].

Photons are not only a vehicle for conveying a reality, though, but also generate matter by interactions, as already described[87].

Some physicists believe that, through a conscious observer, reality in a macroscopic quantum world is at any moment being "connected". Quantum physicists have discussed the related problems in the context of the so-called **measuring-problem**[88]. Each consciousness-containing system would thus be in a position to produce "matter-containing reality".

What happens to the matter-containing reality, when, for example, consciousness of a human observer during deep sleep, anesthesia, or is "off" otherwisely? Does reality disappear at this moment? Einstein in any case refused on principle grounds to accept such mind games: for him the moon existed, even when he would not look at it.

Is it really so that, for example in an individual deep sleep situation, the matter-containing reality does not disappear, but remains the same? Does it not therefore have to be something, that dynamically maintains the interaction between the factual-real world and the metric-free vacuum during individual "shutdown"?

Immediately the question comes up, whether consciousness-containing-systems must necessarily be linked with any living being?

Theoretically, it is possible that all that is alive, simultaneously shuts down consciousness. In that theoretical case,

86 Wrobel, Sedlacek (2014) *Quantenstaub* - Quantum dust, p 99, p 105
87 dto, American physicist firstly created matter of pure light, p 77
88 J. v. Neumann, *Mathematical Foundations quantum mechanics,* Springer (1932, 1968, 1996).

the matter-containing reality should disappear. But this is hardly imaginable. It is conceivable, though, that there must be something outside of the living consciousness-containing systems, which maintains reality at any moment.

The assumption lies thus close that there are - next to conscious processes of life - other unknown processes that can switch matter-containing reality. Or, the elementary process of consciousness is actually a universal property of all, that exists as a substance, and is not bound to liveness.

At a quantum level it is to assume that, wherever the interactions depend on mere, objective quantum chance, the criteria of elementary consciousness are met. The latter presentation describes better the actual conditions on emerging reality, but on the cost of the consequence that non-living, including something dead, too, contains at least elementary consciousness.

Immediately, notional discomfort creeps back at the idea of a boulder possessing a kind of consciousness. With the existence of elementary consciousness in all inorganic matter, however, an option opens, which explains sufficiently well, why the matter-containing reality would even exist, if consciousness was disabled in all living beings.

4. Evolutionary principles of the development of consciousness

4.1 Consciousness at the beginning of the evolution

Archaea Haloferax volcanii behaves in the replication differently than expected

"It is a form of life that we previously did not know like this."

Archaea were originally discovered under extreme environmental conditions and can survive both, under extreme maximum and minimum temperatures, in highly saline and in strongly alkaline water.

On a genetic level, archaea are more related to the eukaryotes, thus with us humans, too, rather than to the bacteria.

More at: http://www.nature.-com/nature/journal/v503/n7477/full/nature1 2650.html

Complex community of microorganisms in tyrolean ice cave

Researchers discovered so-called moon-milk in the tyrolean Hundsalm ice cave. Each drop contains only bacteria and fungi, also archaea up to a million per milliliter, that grow and thrive on one of the nutrient poorest patches of Earth.

More at: http://science.orf.at/stories/1742194/

@ KDS:

... Archaea Haloferax volcanii is a form of archaea and is a discovery that refutes the replicon model, from which researchers for 50 years assumed that it was one of the basic requirements for life. Quotation from this work:

> *"We have indeed a form of life in front us, but a form of life that we did not know this way: outwardly, they look like bacteria, but inside they are like us."*

According to the replicon model, all self-reproducing living beings must first copy their DNA before the cell is able to divide. This is done at a variety of origins of replication. Are these origins of replication in eukaryotes - such as in humans, damaged or missing, they prevent replication, which leads inevitably to cell death.

In Haloferax volcanii, the mechanism works differently, though. Here the chain reaction of cell division begins spontaneously, even if the origins of replication were harmed, or are even not available. In the latter case, as shown by the investigations, DNA replication takes place as much as ten percent faster.

It is quite obvious that, at Haloferax volcanii replication, operations run differently from the expected. Also, an unusual energy extraction seems to be possible, as was demonstrated in the archaea found in the tyrolean caves in "moon-milk", in extremely nutrient-poor environment. Archaea manage to stay alive and reproduce, even at places where it seems to be impossible.

Investigators suspect causally a different metabolism, with which methane is produced. By doing so, they perhaps gain life-sustaining energy.

In chapter 3.1 (consciousness as a prerequisite for the evolution of life), a fundamental solution is described, by the use of which life is able to flourish in a hostile environment, in which **consciousness occupies the pole-position**, and, following to that, evolution is triggered.

Ultimately, the common denominator of the three species of archaea, blind cave fish and deep-sea organisms, aimes at "**target-orientated** behaviour". Here we observe quite clearly the will to survive in an otherwise hostile environment. Living beings have to somehow muster emotions, leading to their targeted behaviour. The behaviour does not necessarily have to be objectively "reasonable", but can also be simply subjectively inked.

To behave the way they do, physical resources are provided that are activated for a particular purpose (in the sense of non-deterministic choices between alternative courses of action), which would lead to the term "emotion".

Emotions, however, imply a subjective or phenomenal view of consciousness. Argues lens, it makes sense to use the concept of **need** as an "inclination to pursue a specific goal": as a whole, it is a "goal-directed behaviour for the satisfaction of needs", namely, in the example chosen, to live in a hostile environment.

You described "consciousness and information-processing" as fundamental for the extraction of energy from the metric-free vacuum[89]. Does the previous mentioned: "... to live in hostile environments" indicate what pattern quite different species - listed above - simply live after, or do we have to adopt causal processes of consciousness?

@@@

89 Sedlacek (2012) *Der Widerhall des Urknalls* – Echo of Big Bang, p 151ff

@ NW:

... I am again fascinated by what you found. The discovery of an unexpected DNA replication is in my opinion extremely important. I think, one more building block for a better understanding of the origin of life is found.

And the second item is just as fascinating. The moon-milk seems to contain a complete microcosm. A microcosm in a gated community that requires little, or nothing from the outside to provide for its life.

And finally, I am fascinated about your rationale. I would like to contribute a few additional words to the facts: just like you, I assume that information-processing, already identified as consciousness, is responsible for the capability to exist in a hostile atmosphere, deep in an ice cave, in deep-sea vents, gushing-springs, or other extreme environmental spaces. Without the help of a consciousness-process, it would certainly not be possible to adapt to external conditions, which are drastically different from habitat to habitat, and certainly are not always constant, but subject to up- and downturns.

It is probably fact that the members of the community in their extreme habitats, individually, and in terms of their own needs, do have to make non-deterministic decisions in order to stay alive and to replicate.

Once the criteria for consciousness are obviously satisfied, we can not help but say consciousness processes are essential for the survival of even the lowest levels of life.

In contrast, we discover no kind of brain, nor place in archaea, or other micro-organisms, appearing particularly suitable for the processing of information. We are - like in the double-slit quantum experiment, faced with the question: where does information-processing take place, where is a data store to be found, and so on?

With regard to the purely biochemical processes, such as replication, the necessary information is perhaps saved in the DNA.

But, the recognition of environmental changes requires the **storage of former environmental conditions**. Without the information of a former state, changes cannot be recognized. And, unless changes in its environment are recognized by a living being, it is unable to respond, take decisions, or adapt to it. Without adaptation, it ultimately dies.

But, obviously, the micro-creatures managed to maintain and to adapt over extremely long periods of time to the most extreme habitats. So, where is the current environmental information stored? Where does associated information-processing take place?

We already answered the same question elsewhere and came to the conclusion that it can only be the metric-free vacuum, in which maybe something is stored, in which also at least part of information-processing takes place.

Similarly, we found out that quantum physical processes, entanglement included, play a general role in cells and in the DNA. This can not be different within micro-organisms and archaea at extreme locations.

Archaea, which are at the beginning of the food chain of a community, probably collect energy they need to live, with the help of quantum entanglement processes conducted by photons, which also occur in dark and extreme habitats. This procedure happens probably similar to photosynthesis in the marine algae, which also make use of quantum mechanical entanglement for their processes. Entanglement is required to collect low-energy photons, until the amount of energy for biochemical processes is on a sufficient level.

The required consciousness-processes probably take place to a substantial portion in the metric-free vacuum. Without a permanent connection of the 4-dimensional quantum-world with the nonentity, nothing would work at the micro-organisms; no nutrition, no replication, no adaptation to changing environmental conditions, no sustainment of life.

4.2 Life in eternal darkness

Blind and deaf cave fish discovered

In the depths of most caves constant darkness prevails, many cave animals have therefore reduced their sight. To help them to discover prey, and to find their way, the other remaining senses should supposedly work more accurately, to compensate the lack of light.

But now researchers have discovered for the first time a cave fish that is not only blind, but also almost deaf.

More at:

http://www.scinexx.de/wissen-aktuell-15842-2013-03-27.html

The Mystery of the eyeless fish

For the genetic and evolutionary world, blindness of fish is still much of a mystery, though. Theories and results, there are many of them - certainties, however, are very rare. Even Darwin found himself at the end of his evolutionary knowledge, with regard to the eyeless aquatic animals.

More at:

http://www.spektrum.de/news/fische-n-im-trueben/938375

Blind cave fish as a model for innovative sensors.

Although the blind cave collector (Astyanax fasciatus) sees nothing, it could play a significant role in the development of novel motion sensors.

More at:

http://www.proplanta.de/Agrar-Nachrichten/Wissenschaft/Blinder-Hoehlenfisch-als-Vorlage-fuer-innovative-Sensoren_article1238354905.html

Deep- sea: life in absolute darkness

In the deep-sea, there is neither sunlight, nor are there any plants. The realm of eternal darkness begins 200 meters below sea-level. However, even in the deepest depths, there are big and small living beings.

See more at:

http://www.geo.de/GEOlino/natur/tiefsee-leben-im-dunkeln-1754.html?eid=61573

@ KDS:

...the theme of blind cave fish is not least so exciting because of the fact, that development periods of many millions of years can be overlooked: how do fish get there, and how are they able to survive under such extreme conditions? Is there a look-at-across link to archaea? Perhaps, an evolutionary model could be derived on the basis of quantum-theoretical considerations?

As we have meanwhile learned, slug Elysia timida uses the ingested plastids of marine algae quite differently: either the deposited plastids along the digestive tube serve to produce energy via photosynthesis, or the plastids digest according to requirements - for the case, they became functionless. This happens, for example, when the slug is kept in complete darkness.

Similarly, but in reverse order, deep-sea organisms seem to act: if necessary, they rise from the depths to the surface to get at plants, that may exist from 200 meters up to a maximum depth, in order to meet their energy needs. In this example, two things become very clear:

- Plants explicitly need light to gain energy through photosynthesis, which is a very effective way to generate it. The energy yield is highly efficient, due to quantum effects. Although plant eukaryotes also feature mitochondria, they can not entirely dispense of light, to produce energy.

- Animal eukaryotes, such as Elysia timida, do not necessarily need light to live, but are able to take advantage of this when needed, to gain energy. In contrast, deep-sea organisms seem to exist mere light-independently.

In Earth's history, light-dependent creatures existed probably first[90]. Later, forms developed from them, that could also lead a light-independent life. This begs for the question, why just animal eukaryotes evolved independently of light, although they used photosynthesis, and their energy yield was still very efficient?

The highly advanced photosynthesis of plants must have evolved in some form in animal eukaryotes for effectiveness and efficiency reasons. There are no doubts that for this performance quantum effects can be assumed as basis.

Principally, for their specific life processes and reproduction, two different structures are available: plasmons (extrachromosomal functions) and DNA (chromosomal functions), in which energy is normally supplied by mitochondria.

Is it conceivable, that one of the two structures, or even in both at the same time, photosynthesis similar processes are being effected? In any case, such options to gain energy, could be explained by quantum-based-association even under extreme conditions. Immediately after suggesting/assuming this, the question then arises, however, how quantum effects, such as the case of plants, occur even with animal eukaryotes ? And then, would not the plasmon DNA and original DNA have to have the crucial structures?

@@@

The importance of cyanobacteria for the development of life

Judging by recent discoveries, the cyanobacteria belong to the oldest autotrophic organisms occurring on Earth, as 3.5 billion old representatives of their class were found

90 http://www.mpimp-golm.mpg.de/32235/research_report_383967

already. Some time later, the first organisms developed, the early prokaryotes.

More information at:

http://de.wikibooks.org/wiki/Cyanobakterien:_Bedeutung_der_Cyanobakterien_für_die_Entwicklung_des_irdischen_Lebens

@ NW:

… further above I presented in detail photosynthesis (chapter 2.4: Mode of action of plastids and mitochondria) according to following active principles of quantum entanglement, and the deductive conclusions resulting from them. With regard to answering your questions, the behaviour of cyanobacteria (= blue-green algae) for the development of life on Earth should be particularly helpful: in two respects, they played an important role in the evolution of life on Earth. On the one hand, they were involved instrumentally as pioneer organisms on Earth, helping to enrich the Earth's atmosphere with oxygen, and to make her habitable for complex life. On the other hand, they "helped" to develop endosymbiosis, which was quite essential for the growth of the vegetable and animal eukaryotes.

Do chloroplasts now solely stem from once phagocytosed cyanobacteria? Photoautotrophic[91] eukaryotes always have photoactive cell organelles. They bear witness of the fact that we are able to clearly see, how endosymbionts adapted to the surrounding cell. In comparison with photoactive cell organelles, unicellular algae were followed by red algae, green algae, up to green plants, the prokaryotic nature of the algae be-

91 **Photoautotrophy** is the use of light as an energy source in autotrophy. The light- energy is converted into chemical energy (ATP), to use it for the construction of building and reserve substances from inorganic materials.

comes very clear: in green organelles of plants, chloroplasts were found with chlorophyll pigments, needed for photosynthesis as chlorophyll a and b; cyanobacteria, however, dispose only of the photoactive phycocyanin and phycocrythin.

In marine symbioses probably two things happened: in addition to cyanobacteria more prokaryotes were found in the meantime, that are capable of oxygenic photosynthesis. They are called prochlorophytes and use - as plants do - chlorophyll a and chlorophyll b, but do not have the dyes phycocyanin and phycocrythin. They are prokaryotes, though, and stand next to, or in between chloroplasts and cyanobacteria, as they have characteristics of both.

By answering the question: "what does Elysia timida do, that differs it from other plants?", we find important clues of, how "photosynthesis" could have evolved evolutionarily and, which structures exercise in animal eukaryotes functions, at least rudimentary ones, on the same principle: the endosymbiotic theory is based on the fact that relatively large prokaryotic unicellulars sometimes add smaller prokaryotes to itself, without metabolizing them. While plant eukaryotes endosymbioticly use exclusively plastids (chloroplasts) to live, Elysia timida affords the luxury to decide, to take on the function of photosynthesis for energy production in plastids, or claims these to be achieved, by digestion of the same. So, there are alternative selections. The slug can only consciously make a decision for the use of the two functions, to produce energy.

For a way of life independent of light, there seem to exist - in addition - alternative options:

Bacteria produce methane or oxygen without light [92], for instance.

92 see http://www.spiegel.de/wissenschaft/natur/forscher-entdecktenmikroorganismen- deep-in-the-ocean-earth's crust-a-888635.html

Or: deep-sea organisms are possibly able to convert radioactive material into usable energy[93].

Archaea manage to gain the necessary energy for life (keyword: energy metabolism)[94] without light and without oxygen.

4.3 Ultra-weak photon emission (biophotons) by biological systems?

> ***Photonic Communications and Information Encoding in Biological Systems.***
>
> *Currently, the term 'bio-photons' is attributed to the optical and UV photons emitted by the living bio-systems in the processes which are different from standard chemiluminescence. The photon radiation in the form of short quasiperiodic bursts was observed for fish and frog eggs, hence the communication mechanism can be similar to the exchange of binary encoded data in the computer nets via the noisy channels. The data analysis of fish egg radiation demonstrates that in this case the information encoding is similar to the digit to time analogue algorithm.*
>
> *More at:*
>
> *http://arxiv.org/pdf/1205.4134v1.pdf*

93 see http://www.scinexx.mobi/dossier-detail-66-14.html, see http://www.deutschlandfunk.de/leben-im-inferno.740.de.html?dram:article_id = 206954

94 see http://books.google.de/books?id=u80VCx807kQC&pg=PA263&lpg=PA263&dq=energiestoffwechsel+prokaryoten&source=bl&ots=CoDoq96YYm&sig=gVjMt3XQNPT7biTYUp005H5HM90&hl=de&sa=X&ei=pbXbU9r7BIuw7Aa5vYCYDA&ved=0CEIQ6AEwBA#v=onepage&q=energiestoffwechsel%20prokaryoten&f=false

@ KDS:

... in the present study, the effects of ultra-weak radiation (bio-photons[95]) were detected in fish and frog eggs. Could one not assume that such a specific radiation for the sustainment of life is "artificially" created - in a biological sense, to gain energy analogous to photosynthesis? Is it then not conceivable to apply this principle in a meaningful way, in archaea or multicellular organisms, that live in complete darkness, such as cave fish or deep-sea organisms?

The debate about the existence of ultra-weak radiation, to which the DNA is accepted as a source, goes on for almost a century now. Originally discovered by Gurvich/Gurvich as "**mitogenic radiation**," the term "bio-photons" established itself in the meantime for this type of radiation. One of the protagonists, Popp, assumes a quantum physics-like coherent photon behaviour, much like a laser. Technical developments have emerged from it by now. Up to the present day, however, the basic questions, in particular the “how-” and “why”-questions, have not been asked, yet.

We now made acquaintance with photosynthesis and learnt that it is a very efficient quantum process, involving the plastome, which is used to grow, or sustain life of a plant. The necessary energy is provided by entangled photons. How can we bring to our mind something similar happening in animal eukaryotes, bacteria and archaea?

Archaea also occur in deep rock formations, partially several kilometers below the surface, without the possibility to expose themselves to light. They are single-celled organisms with a generally self-contained DNA molecule (also re-

95 The term is used for those **biophoton** lightquanta which are part of the ultra-weak photon emission (UPE) of biological origin. The radiation differs from bioluminescence by its much lower dimension of by magnitude intensity.

ferred to as "circular chromosome"), which is arranged in a small volume and is referred to in this form as the core equivalent. Archaea obviously belong to prokaryotes. In them, there are no cell organelles, but there are indications that they form a functional, similar-like-filament cytoskeletal to stabilize their shape.

How can you imagine yourself the mechanisms of growth and vitality in archaea? Where do they find the energy they need to sustain life and reproduction?

In 1866, Haeckel speaks in his "Generelle Morphologie – General morphology, engl." of an "elementary psychology"[96].

Thus, unicellular life- forms, for example **protists**[97], similarly express feelings and will, or similar instincts and movements shown by higher organisms, such as the infusorians[98].

Both, the behaviour of these excitable unicellulars to the outside world, as well as many other expressions of life as response (for example, the construction of the housing of rhizopods[99]), convey the impression that, as they do so, this can be interpreted as clear signs of conscious soul-activity.

Assuming the cellular theory of consciousness as characteristics of life for each cell, and facilities of each mental function with an consciousness component, you will need to

96 http://www.biodiversitylibrary.org/item/22319#page/11/mode/1up

97 **Protists** (greek protista, "fundamental beings", "first-fruits") are a group of non-related unspecified microscopic creatures, to which belong all one-only to multi-cellular eukaroynts, such as algae, protozoa and some fungi.

98 **Infusorians** are small animals, which develop i.e. in the infusion of plant material (ciliates).

99 **Rhizopods** are defined as heterotrophic protozoa whose protoplasm can form protuberances for the purpose of feeding or locomotion, called pseudopods (pseudopodia).

have to subscribe to each separate protist a cell-consciousness. The material basis then would be the whole cell-plasma, or the core, or a portion of the same.

Assuming an ultra-weak radiation derived from DNA, the conclusion could be drawn then, that archaea draw power through its circular chromosome. In analogy to the photosynthesis it should be energy, which is taken from quantum mechanical entanglement.

Archaea, just like every other "that-which-is-being" of reality, can be interpreted as macroscopic quantum objects that are equipped with some kind of consciousness. In principle, archaea are well able to elicit such information-processing that helps them to gain energy, necessary for life and reproduction, and, even if only at the quantum level, opt for certain life-sustaining processes in order to "borrow" energy from the metric-free vacuum.

In contrast, plant eukaryotes endosymbioticly collected plastids (plastome) for the function of photosynthesis, to gain energy.

In animal eukaryotes, chondroma could be adopted in the mitochondria as either being analogous to it, or being something different, something in some way rudimentary. The existing biological signal structures must play an important role of recording, or disclosure of information.

And what about cyanobacteria? Cyanobacteria are organisms that occurred at a very early stage of life-development and were able to generate energy for life processes fotoautotrophicly. In the further development history they entered into an endosymbiotic relationship with eukaryotic cells. Now, if thus the basic ability to photosynthesis not only exists in vegetable eukaryotes, but also in animal ones (see behaviour Elysia slug), it is then of main interest what we can deduce from multicellular organisms such as cave fish, deep- sea creatures, or even humans.

If ultra-weak radiation is produced by an organism itself, it is imaginable that this feature mainly developed itself to exist light-autonomously. If structures for the generation of ultra-weak radiation are created - rudimentarily in mitochondria as a power plant of a cell, or elsewhere, in very harsh conditions, maybe similar to a photosynthesis process, light energy is transformed into that amount of chemical energy (ATP), which is necessary for the build-up of vital nutrient reserves from inorganic substances.

The evolutionary progress would consist of the option to lead a life, unlike plants, without light. Acceptance of this viewpoint could help to explain, why some forms of life can even exist in an otherwise hostile environment: they produce the light they need for their biochemical processes themselves.

This raises the following questions:

1. Assuming the case that there is something rudimentary applied in animal eukaryotes, which in principle takes the function of an otherwise light-dependent photosynthesis: could the requisite light then be generated via ultra- weak radiation?
2. Could the ultra-weak radiation serve to activate signal structures in order to initiate, for example, DNA replication?

Also included into these questions is the consideration, what precisely is the rudimentary thing? Could it be that it is the chondroma itself, located in the mitochondria?

@@@

@ NW:

... should the source of bio-photons actually be the DNA, then it is very likely to be a general phenomenon that occurs in all DNA-containing cells. It is known that DNA works the same way everywhere, even in archaea.

Along prevailing view among scientists, radiation is based on known chemical reactions within the metabolism.

To me, it only seems to be a little strange that chemical reactions produce visible light, as, by burning nutrients, process heat usually arises. And the photons of process-heat do - in my view- not belong to the visible light spectrum.

It can therefore not be excluded that bio-photons must have a different source. If the suspected source of bio-photons is DNA, this can also mean that DNA requires a few photons out of the visible spectrum, in order to **trigger the process of replication.** It is in any case well conceivable for me that light is generated to produce biochemical energy from light, through a photosynthesis-like process, to maintain the major processes of life.

I realized that the structural formulas of amino acids, from which the DNA is composed of, contain in many places hydrogen H+. Hydrogen has usually only a single electron. To each possible electron trajectory belongs a specific energy level, that is represented by the sum of potential energy and the kinetic energy of the electron. The electron orbits are numbered from 1 to infinity. The higher the trajectory number is, the farther the electron is away from the nucleus.

Electrons emit visible light when they change their tracks, numbered 7 to 3, to track number 2 (known as Balmer series). When the electrons move from track 5 to 2 to track 1, high-energy UV light is emitted. I think, this is less well-liked by the DNA because of the risk of mutations. When electrons of track 7 to 4 fall back to track 3, then infrared light is emitted (heat).

The question now arises, why do electrons change their track from 7 (up to 3) to 2? There are probably two possibilities:

a. Quantum random/mere chance (because it simply pleases the electron).

b. Because bond-energy may be required for the incorporation of hydrogen into an amino acid. This binding energy is obtained from high-energy electrons by falling to a lower-energy track (track 2).

Regardless, of whether a. or b., or both are true, I think it can be assumed with certainty that the bio-photons actually stem from the operations in the field of DNA. Other places of origin seem to be unlikely.

The next question in this context arises: how do electrons reach the higher-energetic tracks 3 to 7 (the process is called excitation)? There are three options to answer:

a) **Thermal excitation**. The kinetic energy of the electron is increased by the required amount of energy by displacing the hydrogen atom by heat (photons), or process heat from the burning of nutrient in a larger atomic or molecular motion (kinetic energy). Collisions between the atoms raise the electrons to higher orbits.

b) **Photoexcitation.** Interaction of the electron with light, i.e., the electron absorbs a photon with sufficient large energy.

c) **Electrical stimulation.** If electrons hit atoms at high speed, they are excited and lifted to a higher orbit. The electrons could arise, for example, from the beta decay of radioactive material.

Case b (photoexcitation) is probably pointless for those micro-organisms which live in eternal darkness.

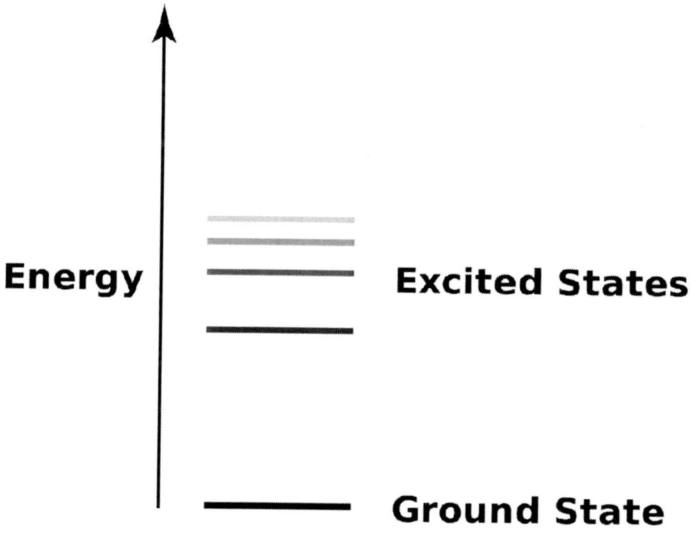

Fig. 7: Energy levels for an electron in an atom: ground state and excited states. After absorbing energy, an electron may jump from the ground state to a higher energy excited state.

4.4 Basic concepts for the wording of a theory of quantum consciousness

@ KDS:

... at this point I want to try a summary of our findings:

Cyanobacteria were one of the first structures that could fotoautotrophicly live as organisms. Through photosynthesis, they could possibly meet part of their energy demand. Pigments, which are rooted in the photosynthetic lamellae system (chloroplast), dispose of chlorophyll b and auxiliary pigments, such as phycocyanin and phcocrythin.

By the way, cyanobacteria - as an outstanding performance, contributed very significantly to the formation of oxygen, and thus to the Earth's atmosphere[100].

In contrast to eukaryotes, DNA of cyanobacteria lies freely in the cytoplasm. Separated cell organelles are absent. Reproduction takes place vegetatively, thus by cell divisions[101].

Like mitochondria, plastids (chloroplasts of plant eukaryotes) possess both, own DNA, own ribosomes and biomembranes as a shell.

Mitochondria and plastids (chloroplasts) of eukaryotic cells developed very early in the course of evolution, by endosymbiosis. In it, prokaryotes, e.g. cyanobacteria and other prochlorophytes - in the case of chloroplasts, and alpha-proteobacterium - in the case of mitochondria, were phagocytosed by other prokaryotes (the origin eukaryote as the predecessor of animal or plant cells). The prokaryotes in the primordial eukaryotes aggrandized in conjunction with

100 http://de.m.wikipedia.org/wiki/Stromatolith
101 http://www.mnf.unigreifswald.de/file admin/Mibi_Oekologie/Cyanobacteria__Brock_Mikrobiologie_.pdf

them, but regressed cell-functionally in the course of time, and took over specific functions in the evolved eukaryotes.

In Short

Plastome (plastids) and **chondroma (mitochondria)** are part of the extrachromosomal system, which is collectively referred to as **plasmon**. Both structures can be regarded as a system with comparable properties:

They have DNA molecules and are responsible for **energy supplies** to maintain life processes. This energy is hence mainly needed to be able to exist - away from the thermodynamic equilibrium, as sine qua no to live at all.

As a common feature, they entered into an endosymbiotic connection. Nevertheless, plastome and chondroma are regarded as two different entities.

Mitochondria often are referred to as the power plant of a cell. Like any plant, they need fuel in order to produce usable energy. The same function can be assigned to the plastids (chloroplasts). Unlike mitochondria, they buy their "fuel" from light, to produce energy.

Now, vegetable, as well as animal eukaryotes have both respectively mitochondria; plastids (chloroplasts) only the vegetable ones, though.

Both eukaryotic species are capable of maintaining life-processes away from thermodynamic equilibrium, due to their "powerhouse function".

They reproduce quite differently: mitochondria multiply by growth and sprouting, plastids (chloroplasts) by division.

In contrast, the overall reproduction, or the multiplication of the cell is delimited by cell division.

The plastids´ power-plant of cyanobacteria possibly supplies energy exclusively from light to sustain life, the vegetable eukaryotes, in contrast, not necessarily, as they possess additional mitochondria. Ultimately, the latter seem to be dependent on light, though, as, at depths below 200 m, there are no plants found, because it is too dark there.

How is the success of the evolutionary process of cyanobacteria from the vegetable to the animal eukaryotes measurable, then? It is probably the ability of light-independent existence of the animal eukaryotes, to which humans count, too. The presence of ultra-weak radiation was detected within the simplest plants, animals, and within humans (Popp: bio-photons). During measurements it was found that the number of emitted photons is linked to the position of the organism on the evolutionary scale: the more complex the organism, the less photons are emitted.

While in simple plants, or animals, an emission rate of 100 photons/cm^2/sec in the wavelength range of 200-800 nm could be measured, the rate in humans is at 10 photons/cm^2/sec (Ruth, Popp: Experimental study of ultra-weak photon emission in biological systems). As emission source, the DNA of the examined living organisms is assumed.

At this point, the fundamental question arises concerning the function of ultra-weak radiation and, especially, its place of origin.

First of all, ultra-weak radiation is not necessarily needed by larger plants, because they possess plastids and can practise photosynthesis with the there existing chloroplasts. Perhaps, ultra-weak radiation becomes relevant in the dark, to maintain the process of photosynthesis at low levels. The

DNA-containing plastid genome of plants is found in plastids, with the photosynthesis pigments chlorophyll a and b.

It should be quite similar in cyanobacteria with their chlorophyll analogies. A particularly high photon emission-rate was verified experimentally in simpler plants. As the birthplace of the bio-photon radiation in cynobacteria, only the chlorophyll-analogues phycocyanin and phycocrythin would come into consideration - the plastome being present in the cytoplasm, because only there DNA molecules can be found.

If in these structures the detection of emitting bio-photons is achieved, another emission source is considered to be located in the mitochondria of vegetable and animal eukaryotes, because more DNA is only to be found in local chondroma.

Plastome (plastids/chloroplasts) and chondroma (mitochondria) are denoted in an umbrella term as plasmon.

The biophoton emission could in any case assigned to a useful function: it serves as a "fuel supplier" for the power plants mitochondrium, respectively plastid /chloroplast, to provide useful chemical energy to maintain life processes, far from the thermodynamic equilibrium.

In summary, a picture emerges, along to which the plasmon, i.e. the extrachromosomal part of the genome, could be perceived as a source of bio-photon radiation.

And how are the bio-photons created?

For the more advanced eukaryotes, it may be theoretically conceivable that a yet unknown process can be held responsible for the generation, but not in cyanobacteria.

The example of stromatolites[102] adumbrates, under which adverse conditions they managed, not only to survive successfully, but at the same time - as a very special performance, to produce oxygen, and thus the atmosphere of our planet Earth.

The obvious idea urges that only a quantum physical process made this effort performable. Furthermore, it can be assumed that information-processing through cyanobacteria again meet the criteria for consciousness (see also Haeckel: Crystal souls: studies on the inorganic life). I want to spare it to myself, to analyze the behaviour of cyanobacteria in detail, to map it to the relevant criteria for consciousness, because the basic explanatory statements have already been made further above.

Should we fail proof of a bio-photon radiation in cyanobacteria, the hypothesis, that in the simplest algae [103] quantum-physical processes trigger an effective and efficient photosynthesis, is refused in terms of a falsification.

As we already know, replication processes in archaea run different from what is expected. They are very successful creatures and have sucessfully enforced in an otherwise hostile environment, and are able to exist light-independently.

Archaea as unicellular organisms have usually a self-contained DNA molecule (circular chromosome), and no cell organelles. Accordingly, they neither have plastome, nor chondroma. Their powerplant could be represented by the

---

102 **Stromatolites** are biogenic sedimentary rocks that have arisen as a result of growth and metabolism of micro-organisms in an aquatic environment

103 **Falsification** proves the invalidity of a statement. A scientific theory must be principally falsifiable . Otherwise it is no scientific theory. The concepts of the wording (Foundations for the formulation) of a theory of quantum consciousness are successfully enforced.

plasma itself, or by plasma structures. Their energy-gaining seems to run differently in extremely nutrient-poor environment (see tyrolean caves, moon-milk). In any case, research on bio-photons would be extremely useful for the detection of ultra-weak radiation in them, and would help to explain their differentness.

Additional questions, or rather commentaries:

1. A first step would be the identification of the quantum physical processes, according to which principle energy is supplied to the plasmon system. There are studies, in which mutations in DNA could be detected through "proton tunneling". Is it conceivable, too, that such tunneling effects can also be assumed, i.e. at plasmons?

2. In nanosciences, quantum effects - among others, are studied at carbon nanotubes, by so-called plasmon techniques. A comparison with the plasmons in biology, respectively the plasmons in physics, suggests itself. Can the plasmon technology be regarded as the inorganic physics for the detection of quantum effects?

3. In cyanobacteria, chlorophyll and hemoglobin chlorophyll analogies are common, to oxidize divalent iron to trivalent for O_2-forming. Divalent iron ions can also be oxidized to ferric iron without dioxygen, upon exposure to ultraviolet radiation and blue light
($2\ Fe^{2+} + 2\ H^{+} \rightarrow 2\ Fe^{3+} + H_2$).
Iron ore bands could thus be created by exposure of Fe^{2+}-ions without dioxygen, dioxygen needs not to have to be present at that time. With regard to the dialogue-underlying, original question: "What is

disease?", evolutionary and physiological processes have to be involved in each case, and in all further considerations[104].

4. Information-processing, adopted as an interaction between consciousness of a quantum object and the metric-free vacuum, resembles the principles of cloud technology. For example, a speech recognition system: by speaking into the microphone of a smartphone, information-processing is performed by a programme in the cloud, and as a result – namely the written word – becomes readable on the display.

@@@

@ NW:

… **proton tunneling** is, just as electron tunneling, which led already to many technical applications, a general quantum physical effect - which certainly occurs not only at any particular DNA, in which the detection was carried out, but was also found in biology in similar molecules, especially in plasmon structures.

Quantum mechanics tell us in addition that proton tunneling occurs along with all atoms, that it happens everywhere, in most atoms in a very low abundance, though.

With regard to your comment in 4: I think you found a vivid metaphor for this specific process.

@@@

104 Wrobel, Sedlacek (2014) *Quantenstaub* - Quantum dust, p 33

@ KDS:

... in my opinion, the previous findings suffice to formulate a theory of quantum consciousness - due to evolutionary development processes, based on scientific criteria[105]: if an ultra-weak luminescence in organic and inorganic substances could be detected, then the theory would be confirmed. There is also a possibility to falsify it (see page 112).

The issue of "inorganic substance" in this context presents still an unaltered problem to me. Is there an evolutionarily developed quantum consciousness in inorganic substance, too?

In 1870, Ewald Hering labelled "memory as a general function of organized matter", and high-lighted the importance of this soul activity, "to which we owe almost everything, what we are and what we have".

Haeckel took up Hering's thoughts and, in 1876, developed, 'Die Perigenisis der Plastidule (plastidules = plasma molecules), or wave-procreation of the vital particles; an attempt to a mechanical explanation of elementary processes of development". He wanted to prove the "unconscious mind" to be a general, all-important function of all plastidules[106].

Only living plastidules, as the individual molecular groups of the active plasma, are reproductive and thus have a memory; this is the main difference between the organic nature to the inorganic one. Hence, the following is allowed to say: "heredity is the memory of plastidules".

The elementary memory of unicellular protists is composed of the molecular memory of plastidules, from which

105 Wrobel, Sedlacek (2014) *Quantenstaub* - Quantum dust, p 158

106 Hypothetical molecules or molecule groups, which are designated by Naegeli as micelles, other than bioplasts

its living cell-body is build up. Probably, no further facts for the amazing benefits of unconscious memory in these unicellular protists are more informative, than the infinitely varied and regular formation of their protective equipment; the shells and skeletons (Haeckel: Weltwunder, Wonders, engl., Scale of consciousness - cell-consciousness).

The statements of Hering and Haeckel can be interpreted as an intellectual foundation, which I “labelled” the "evolutionary theory of quantum consciousness" (see chapter 5). The analogy of the approaches of both scientists is amazing. Even though the assumingly morphogenetic approach was more important to Haeckel, of both can be presumed that they recognized consciousness as something elementary and, both referred to the quantum theory, even though they did not even have approximately knowledge of anything like it.

Furthermore, Haeckel stated that consciousness was, like any other mental activity, a natural phenomenon and thus subjected to the law of substance. In comparison with different theories of consciousness, he specified the "atomistic theory of consciousness" to be an elementary property of all atoms. It probably originated in the main in the difficulty, some philosophers and biologists had with the question of the primordial development consciousness. The scientists of those days hypothesized on the elementary property of all matter, equalling to the mass attraction, or chemical affinity (= chemische Wahlverwandtschaft). Thus, each chemical element had an individual consciousness (Haeckel. Weltwunder – Wonders, engl.). Consciousness of the soul, Atomistic theory of consciousness).

At first sight, the reader is inclined to regard the atomic theory as a possible basis of the quantum theory. At second sight, though, Haeckel, in his dispute with E .Du Bois-Reymond, distanced himself a quantum from it:

> *„ In my 'Perigenisis der Plastidule' I quite clearly emphasized that I only in my subconsciousness attributed the imagined elementary psychic activity of sensation and volition to the atoms."*

At this point, I would like to summarise the further above discussed issues one more time (see chapter 3.5, studies and thoughts on inorganic life).

> *On the quantum level one can assume that - wherever interactions depend on mere, objective quantum chance, the criteria of an elementary consciousness are met, too.*

This, however, leads to the consequence that inanimate things, dead-whatsoever included, must have a quantum consciousness. And immediately discomfort creeps notionally over me at the idea of a boulder, having something like consciousness, albeit at the lowest stage.

In the same chapter I implemented the statement, *"... why matter-containing reality would still exist, if consciousness of all organic matter (= everything vivid) was shut down."*

So, was Einstein right with his assumption, that the moon existed, even though he would not look at it?

@@@

@ NW:

… referring to your last question, I would like to answer as follows:

Assuming, the decoherence theory is right, the moon exists, even if Einstein is not looking at it. The decoherence theory only requires that the moon enters into any interaction. Interaction happens solely because of the fact that the sun bombards it with photons.

4.5 Ultra-weak photon emission as a universal phenomenon

Interaction of bacterial cells with weak light emission from culture media

Chemiluminescence can be observed from almost all objects being subject to oxidation reactions. Liquid culture media as they are being used in microbiology may show an emission intensity of up to 105 photons / min cm^3 under aerobic conditions, depending on preparation and composition. We report on a very efficient interaction between bacterial cells of a wide variety of strains and the mechanisms involved in the generation of this ultraweak photon emission, leading to an almost complete elimination of the light emission. This new phenomenon seems to be a common behaviour, that can be observed with eukaryotic cells and strains of the Archaea, too, and may be correlated to the varying oxygen tolerance of the strains employed. It is shown that the cell membrane is probably involved in the elimination of the light emission, hinting to the existence of active centers in the membrane.

More:
https://bioorganische-chemie.uni-hohenheim.de/publikation/interaction-of-bacterial-cells-with-weak-light-emission-from-culture-media

@KDS:

... I found a publication that provides evidence of the existence of ultra-weak photon emission at archaea, bacteria and eukaryotes, too:

Linoleic Acid-Induced Ultra-Weak Photon Emission from Chlamydomonas reinhardtii as a Tool for Monitoring of Lipid Peroxidation in the Cell Membranes.

Reactive oxygen species formed as a response to various abiotic and biotic stresses cause an oxidative damage of cellular component such are lipids, proteins and nucleic acids. Lipid peroxidation is considered as one of the major processes responsible for the oxidative damage of the polyunsaturated fatty acid in the cell membranes. Various methods such as a loss of polyunsaturated fatty acids, amount of the primary and the secondary products are used to monitor the level of lipid peroxidation. To investigate the use of ultra-weak photon emission as a non-invasive tool for monitoring of lipid peroxidation, the involvement of lipid peroxidation in ultra-weak photon emission was studied in the unicellular green alga Chlamydomonas reinhardtii.

More:
http://www.plosone.org/article/info%3Adoi%2F10.1371%2Fjournal.pone.0022345

For cyanobacteria, evidence also exists, too...

Measurements of ultra-weak photon fluxes of biological and non-biological materials

Sensors for the measuring of the ultra-weak photon radiation can be used for the investigation of surface radiation of biological and non-biological materials.

The emitted radiation from matter, and the way, how implied-absorbed and re-emitted radiation takes place provides information about material properties, material conditions, interaction processes, and how changes are made.

*The expected irradiance is often less than 100 photons/s*cm^2, and the radiation intensity is in the range of 10^{-19} W. The detection limit of conventional radiation receiver is irradiances between 10^{-16} 10^{-8} W/cm^2.*

More:
https://www.tu-ilmenau.de/fileadmin/public/lichttechnik/Publikationen/1999/Bieske99.PDF

... and just as well for ordinary matter, as the last article shows.

For those reasons, I am furnished with strong information for the evidence/verification of my theory of quantum consciousness.

@@@

@ NW:

... interesting experimental work, above the ordinary, especially those from Ilmenau.

This means nothing else than everything (!), indifferent of biological or non-biological matter, emits ultra-weak photon radiation, with wavelengths that are part of the visible light region. And, depending on wavelength and counting rates (cps), conclusions can be drawn on the material (-> information):

> *"This radiation has been known to correlate with various biological processes, for example, with DNA multiplication, cell division, photosynthesis, photo repair processes, response to external environmental influences, diseases, and the passing away of the organism. "*

... and:

> *"In the field of non-organic materials, it is also conceivable, that [...] allow to conclude on structure and material properties. By doing so the observation of processes (such as the ageing of the material) is made possible.*

And one more comment: the emission of ultra-weak photon radiation is quite possibly a consequence of the quantum tunneling effect, even if nothing of it is written in the article.

@@@

@ KDS:

... yes, it is quite obvious the proof of a universal principle in the ultra-weak radiation. I think we can emanate from the following interaction:

- **Macro level.** Microwave radiation in the universe and interaction with the cosmological background field.

- **Meso level.** Interaction of photons of different wavelengths with vegetable and animal eukaryotes (animals, plants).

- **Micro level.** Interaction by photons of different wavelengths with unicellular cyanobacteria or archaea...

... at which elementary particles can be considered as a condensation of radiation of different wavelengths.

It is probably also true that in universe there is a weak radiation of different wavelengths (not just microwave radiation), which has either not yet been measured, or is not commonly reported.

Now, only one more ingredient is missing: what is it that gives form, respectively gestalt, a structure?

It can actually only be the influence of the gravitational force. I already asked this provocative question, because to me there are striking analogies between entanglement and gravity in a similar form[107]: is gravity mediated through entanglement of photons (see also section 2.5 Coherence of gravity, entanglement and life processes)?

4.6 How consciousness can be demonstrated experimentally in bacteria

@NW:

... after behavioural psychologists were able, to establish consciousness in animals such as chimpanzees, elephants or ravens, we also should be able - along to our theory, to demonstrate experimentally consciousness processes in prokaryotes (bacteria and archaea).

Primary consciousness is, as already discussed, a simple form of consciousness which is roughly comparable with the

107 Wrobel, Sedlacek (2014) *Quantenstaub* - Quantum dust, p 50

functions of the subconscious. It does not include the self- or ego-consciousness, which we know of us humans. Consciousness is information-processing and serves an animal to adapt to new requirements, or changes in external circumstances. If an animal decides for an undetermined manner between possible alternatives, and the decision serves to satisfy its needs, then you can at least emanate from primary consciousness.[108] On the other hand, primary consciousness can not be assumed, if actions are entirely automatic, or are random response to environmental stimuli, and show no decisions between alternatives.

Prokaryotes have flagella to move around by swimming. They only benefit from their mobility when they realize, where to swim. From their orientation reaction (taxis), that is their orientation to a stimulus, or an environmental factor, information-processing can be concluded, which is a prerequisite for consciousness. We distinguish the stimulus-related responses and the directed-away stimulus, from avoided or startle response (negative taxis).

Chemotaxis orientates itself, for example, by targeting on the concentration of a substance. Aerotaxis is the orientation to oxygen. This is a special form of chemotaxis or energy taxis. Phototaxis is the focus on the brightness and colour of light, and galvanotaxis is the orientation along electric fields, just to mention a few taxis species. In the internet, there is a short video of the paramecium, as it orientates itself along an electric field[109].

Many bacteria among prokaryotes are able to simultaneously detect the concentration of food substances, oxygen, or light, and to accordingly govern themselves. As long as they do not recognize food substances, for example, they swim for a while

108 For definition of consciousness see box on page 66
109 http://youtu.be/-U9G0Xhp3Iw

in a random direction, and then change direction, to swim again next into a different direction, for a while. With decreasing amount of concentration, they frequently change direction. With increasing concentration, however, they swim again more purposefully to a place of higher concentration. They show the same behaviour with respect to the oxygen concentration and light[110].

From their behaviour can be deduced that the bacteria are able to timely resolve, whether the concentration becomes stronger, or less stronger. They hence assert **changes** in environmental conditions by storing a previous state, in any way. This sole reason allows to state the existence of information-processing. The bacteria also show a need (= inclination to pursue a goal), to swim to the place of higher food and oxygen concentration.

It can also happen, though, that two different needs are not compatible with each other. For example, the higher oxygen concentration can be opposite to the place of higher food concentration. From this fact an experiment can be construed, which allows conclusions, whether bacteria show primary consciousness, or not.

If there is an area between the two places, in which only the respective need is satisfied, and if there is an area, which provides a stronger stimulus, and the evaluation, which of both stimuli is stronger fails, then the there located microbe must decide, which stimulus it pursues, i.e., the location where to swim. An alternative would be: it remains disproportionately long in the range of equally strong stimulus-assessment, as the microbe is unable to decide.

Pre-tests help to identify both, how large the area of the same stimulus-intensity is, and how the time frame is for the approach to a place, to supply a want. If, during the main test,

110 Cypionka (2006), pp 33f

the microbes in majority clearly exceed this time frame, you then know that the microbes can not decide. Failing this means conscious decisions when, at the same time, the other criteria for consciousness are met.

If the decision - in majority, fails within the time frame identified in the pre-test, it can not be a determined one, as the experiment was designed in a way that the strength of the stimuli for the microcrobe was appraised equally. Then we are dealing with a non-deterministic choice between alternative courses of action. It is the decision to swim to either way to satisfy a need.

In summary, the behaviour of microbes permits perception of information-processing to either a non-deterministic choice between alternative courses due to changes in concentration of various substances, or to environmental conditions, which lead to a purposeful behaviour in favour of the satisfaction of needs. Or, there is an accidental approximation to one of the possible places – in a more classic sense – to exceed time frames, determined in pre-tests. From the experiment results can be concluded, whether microbes met criteria for consciousness, and thus provide primary consciousness.

5. The theory of the evolutionary quantum consciousness

5.1 Self-consciousness, primary consciousness, elementary consciousness

@ KDS:

... I would like to engross now the topic and finalize:

Sir Karl Popper elaborates in his book: "*Alles Leben ist Problemlösen* - All life is problem solving, engl.", chapter: 'Science teaching in developing theoretical and logical point of view' (1972)..., that

> *"the realistic world view along with the idea of approaching the truth, seems to be indispensable for an understanding of the increasingly idealized science",*

... and asks himself the question, "where the crucial difference between the amoeba and Einstein is", and answers it as follows: the amoeba flees from falsification. Its expectation is a part of itself, and pre-scientific carrier of expectations, or hypotheses are often destroyed by refutation of the hypothesis. Einstein, on the other hand, objectified his hypothesis, which is, however, outside of him. The scientist can destroy his hypothesis by his own criticism, without being crushed by it. (Popper: All life is problem solving).

So, if e.g., the amoeba wants to try obtain food that lies behind an obstacle, it must make a decision within the meaning of "inclination to pursue a particular goal". It opts for a test-movement in terms of a hypothesis, to by-pass the obstacle and to thereby gain access to the food.

In case the hypothesis is wrong, amoeba can meet with a fatal end, if it starved to death. Hence, in the real world, the amoeba flees from falsification to survive. Einstein, however, survives any falsification in an abstract world and, on the contrary, can increase his scientific fame.

> *Obviously amoebae have a consciousness that allows them to - not only solve problems, but also to survive in the real world.*

Is this consciousness an ego- or self-consciousness, or is it a "primary" consciousness, which would allow a living macroscopic quantum object to exist in a quantum world as a self-organizing, dissipative nonequilibrium system - far from the thermodynamic equilibrium? And can of a boulder in a thermodynamic equilibrium assumed in the same manner, that it has a primary consciousness?

Haeckel in his dispute with E. Du Bois-Reymond did not use his hypothesis of "atomic consciousness" aggressively. "In my 'Perigenisis der Plastidule', I explicitly emphasized that I imagine the elementary psychic activity of sensation and will, that can be ascribed to the atoms, only unconsciously". (Haeckel: *Welträtsel* - Wonders, , engl.).

What could a primary-like consciousness of anything inorganic look like, which reached thermodynamic equilibrium and is therefore dead?

Photons, electrons, and also larger (carbon) molecules (fullerenes) make by passing the columns in the double-slit experiment a non-determined decision, …

... which means, they decide individually what they do - as an inclination, to pursue a certain goal - namely, to pass one or both columns and, somehow seek no predetermined location on the observation screen, to become factually-real at the impingement location.

The more quantum objects strike, the more the big picture on the observation screen resembles a wave pattern (interference pattern). Is alternately one of the gaps closed, the quantum objects recognize in time, and opt for other places on the observation screen - the result of the overall picture no longer looks like wave pattern.

The decision, which place they intend to visit, is obviously immediately made - instantly - at the moment of passage, as a result of information-processing.

Because of the behaviour of photons, elementary particles and molecular structures, one can imply a kind of "basic or elementary consciousness"[111].

Ultimately, the behaviour of quantum objects in the double-slit experiment means: both, organic and inorganic substances have **elementary consciousness**. This must be distinguished from "ego- or self-consciousness" as a highly developed form of consciousness, as occurs approximately in all humans and some primates, and the "primary consciousness", that can be well detected in all animate beings. The processes of consciousness of the subconscious are part of primary consciousness.

The primary form of consciousness can only be assigned to macroscopic quantum objects that have not yet reached the thermodynamic equilibrium, which are, along our speech and way of thinking (syntax, semantics) "alive". Live-

111 Sedlacek (2012) *Widerhall*, Reverberation, p 64, 73, 95ff and 168 ff

liness in this context means the attempt to survive between the two extremes: heat and cold [112].

This means, Haeckel`s malaise can be explained sufficiently well: his more restrained assessment of an "atomistic theory of consciousness" probably originated mainly from the difficulty some philosophers and biologists of his time had with the question, on the first emergence of consciousness.

Elementary property of all matter

Scientists from Haeckel's time attended an elementary property of all matter, equal to the mass attraction, or chemical affinity. Thus, each chemical element would have an individual consciousness.

The adoption of the scientists of that time can be reconciled with our book, we call it ***elementary consciousness****.*

In contrast, Haeckel's idea of consciousness as "the elementary psychic activity of sensation and voilition", which had preferably been subject to the protists, rather associates with, what we call **primary consciousness**.

112 Wrobel, Sedlacek (2014) *Quantenstaub* - Quantum dust, p.20

Summary: Theory of quantum consciousness

- Evolutionary consciousness includes both organic and inorganic substances. It interacts with the metric-free vacuum (= nothing) through information-processing, creating matter-containing reality.
- Herein the theory is included that cosmological evolution is an evolution of the elementary consciousness, whilst primary and ego- or self-consciousness may represent a biological evolution. To what extent this can interact with the metric-free vacuum, is subject of further investigations.
- Selection in the context of evolution follows information-processing triggered by elementary consciousness.
- "Living" macroscopic quantum objects have a differentiated approach over ego- and self-consciousness, primary consciousness, or elementary consciousness.
- "Dead" quantum objects, however, have only elementary consciousness.
- "Vitality" is produced by auto-catalysis, or self-organization and follows information-processing, triggered by elementary consciousness.

- A fundamental falsification could be achieved by the fact that evolutionary quantum consciousness is modelled by a system of mathematical rules. The rules would generally permit verifiable predictions.

5.2 The new worldview

The cosmological evolution quite clearly traces the aim to link something abstract with the real 4-dimensional world. How is this done in detail?

„Preparation“, commonly done in physics, will detach something from the classical physical world in order to deal with this in quantum physics, then. How could the interplay between the metric-free vacuum (= nothingness) and the real physical nature be made comprehensible?

At the beginning stands the mere, objective chance, influenced by nothing at all. Through fluctuation, condensed electromagnetic radiation reaches our world from the metric-free vacuum and becomes real by interaction and decoherence. This process is not reversible in the 4-dimensional world. What life has made of it, is awesome: so, evolution has produced, for example, antennas in vegetable eukaryotes to collect electromagnetic radiation (= photons) with the help of quantum entanglement, until the thus gained real energy is sufficient for the execution of essential life processes, far from the thermodynamic equilibrium.

One might assume now, that everything that has become real, is subsequently subject to the principly pre-determinable random of classical physics: through self-organisation (deterministic chaos) it should be therefore possible, amidst this chaos, to create order and stable conditions. Indeed, nature seems, however, to have found something better: it decides for itself through processes that can be described as deliberate, as it evolves. Evolution can therefore be seen as a conscious self-assembly process. Both, objective chance and conscious self-organization form a type of symbiotic relationship.

The theory of evolutionary quantum consciousness aims at the unifying mutual interdependence: quantum consciousness is **the mechanism, which has the abstract world interact immanentely and instantaneously with the factual world,** in other words: everything that became real within our 4-dimensional world.

Anyway, it seems to be fact that all "that which is being", organical as well as inorganical, alive as well as dead, is capable to connect to the abstract outside of the 4-dimensional world. This notion may at the first moment cause extreme discomfort: it means, seen differently, that information-processing, to which we refer as elementary consciousness, as it shows characteristics of consciousness, creates something at the very moment of interaction, what we know as matter-containing reality.

In fact, an immanent hybrid function seems to be characteristic of our macroscopic quantum world: to generate the matter-containing reality that exists in our 4-dimensional world - energy is required, which is borrowed from the metric-free vacuum. This loan is just there as a sort of promissory note, i.e. "saved" in a "database" and "listed“ as abstract information.

By the thought experiment "Maxwell's demon"[113] it is well known that this promissory note is a perpetual motion machine in our spacetime world, as long as it is not due. It would become payable, when the information in the metric-free vacuum was deleted. By the work of Bennett and Szilard we know, deletion of information consumes energy, but not the original memories[114]. Exactly this fact is our luck,

113 Sedlacek (2012) *Widerhall* - Reverberation, p 89ff

114 Sedlacek (2012) *Widerhall des Urknalls* – The Echo of the BIG Bang, p

though: because the metric-free vacuum is an entity without time and without space, the information from the perspective of our space-time will never be deleted - to it we owe our existence, until further notice.

It is the time-dependent burden, however, we have to carry as physical space-time constructions: some day we finally reach the thermodynamic equilibrium and die. Dead, however, does not mean "shut down" of information-processing, but only the temporal end within of our 4-dimensional world. But this also means that an interaction of any kind, whatsoever, made by, or in the metric-free vacuum, persists[115].

To sum up our new worldview:

There are wired laws that really control our world. They are the laws of an abstract quantum world, in which all that which is being, contributes to create our known matter-containing reality.

The link of the otherworld abstract with the this – world-factual-reality of our 4-dimensional world, requires a basic physical process.

It is a process that already presupposes an evolutionarily evolving consciousness on an elementary level.

101

115 Wrobel, Sedlacek (2014) *Leben aus Quantenstaub* – Life out of quantum dust, p.111

6. Literature

Cypionka H.: *Grundlagen der Mikrobiologie*, 3. Aufl., Springer (2006)

Darwin C: The *Descent of man* (1879), Penguin Books (2004), ISBN-13 978-0-140-43631-0

Dawkins, Marian Stamp: *Die Entdeckung des tierischen Bewusstseins*; Spektrum, Heidelberg (1994)

Dessauer F: *Quantenbiologie*, Springerverlag Berlin Göttingen Heidelberg (1954)

Duve CR de: *Ursprung des Lebens. Präbiotische Evolution und die Entstehung der Zelle.* Spektrum Akademischer Verlag, Heidelberg Berlin Oxford (1994), ISBN 3-86025-187-2

Eigen M, Schuster P: *The Hypercycle - A Principle of Natural Self-Organization*, Springer Verlag, Berlin, (1979), ISBN 0-387- 09293-5

Eigen M: *Molekulare Selbstorganisation und Evolution (Self organization of matter and the evolution of biological macro molecules.)*, Naturwissenschaften Bd. 58(10), S. 465 - 523, (1971)

Gödel K: *Über formal unentscheidbare Sätze der Prinicipia mathematica und verwandter Systeme. Teil I.* In: Monatshefte für Mathematk und Physik 38, S.173-198, (1931)

Görnitz T, Görnitz B: *Die Evolution des Geistigen*, Vandenhoeck & Ruprecht Göttingen (2009), ISBN 789-3-525-56717-3

Görnitz T, Görnitz B: *Der kreative Kosmos; Geist und Materie aus Quanteninformation,* Spektrum Akademischer Verlag Heidelberg (2002), ISBN 978-3-827-41368-0

Görnitz T: *Carl Friedrich v. Weizsäcker – ein Denker an der Schwelle zum neuen Jahrtausend.* Herder, Freiburg (1992), ISBN 3-451-04125-1

Gurwitsch AG, Gurwitsch LD: *Die mitogenetische Strahlung*, VEB Gustav Fischer Verlag, Jena (1959), Lizenznr. 261

Haken H: *Synergetik, eine Einführung: Nicht-Gleichgewichts-Phasenübergänge und Selbstorganisation in Physik, Chemie und Biologie,* Springer Berlin (1977), 3. Aufl. ISBN 978-3540516927

Haeckel E: *Kristallseelen. Studien über das anorganische Leben,* Alfred Kröner Verlag, Leipzig (1917)

Haeckel E: *Die Welträtsel,* Alfred Kröner Verlag, Leipzig (1899)

Haeckel E: *Die Perigenisis der Plastidule. In: Gesammelte Populäre Vorträge aus dem Gebiet der Entwicklungslehre, zweites Heft,* Verlag Emil Strauss, Bonn (1879)

Haeckel E: *Natürliche Schöpfungsgeschichte,* Verlag von Georg Reimer, Berlin (1870)

Haeckel E: *Generelle Morphologie der Organismen. Allgemeine Grundzüge der organischen Formen-Wissenschaft, mechanisch begründet durch die von Charles Darwin reformierte Descendenz-Theorie.* Georg Reimer, Berlin (1866)

Hagemann, R: *Plasmatische Vererbung,* VEB Gustav Fischer Verlag Jena, Lizenznr 261 215/48/64

Kauffmann, S: *Der Öltropfen im Wasser,* Piper Verlag München (1996), ISBN 3492035493

Neumann, J. v., *Mathematische Grundlagen der Quantenmechanik*, Springer (1932, 1968, 1996).

Oehlkers F: *Das Leben der Gewächse,* Springer Verlag, Berlin Göttingen Heidelberg (1956)

Planck, M: *Religion und Naturwissenschaft. In: Physik und Transzendenz,* Hans-Peter Dürr (Hrsg.), Fischer Scherz Verlag (1989), ISBN- 13: 978-3502191704

Popp FA: *Biophotonen* (1983), Karl F. Haug Verlag, Stuttgart (2006), ISBN 987-3-8304-7267-4

Popper K: *Alles Leben ist Problemlösen,* Piper Verlag, München (1996), ISBN 3-492-22300-1, S. 25 ff

Ruth B, Popp FA: *Experimentelle Untersuchungen zur ultraschwachen Photonenemission biologischer Systeme,* Zeitschrift für Naturforschung, 31c, S. 741-745, (1976)

Schimper AFW: *Über die Entwicklung der Chlorophyllkörner und Farbkörper,* Bot. Z. Bd. 41,S. 102-113 (1883)

Sedlacek KD: *Der Widerhall des Urknalls,* Norderstedt (2012), ISBN 978-3-8482-1225-5

Sedlacek KD: *Supervereinigung: Wie aus nichts alles entsteht. Ansatz einer großen einheitlichen Feldtheorie,* Norderstedt (2010)

Wächtershäuser G: *Before enzymes and templates: Theorie of surface metabolism,* Microbiol. Mol. Biol. Rev. 52 (4), S. 452–84, (1988)

Weizsäcker CF von: Der Mensch in seiner Geschichte. Hanser, München (1991), ISBN 3-446-16361-1

Wrobel N, Sedlacek KD: *Leben aus Quantenstaub,* Norderstedt (2014), ISBN 987-3-7357-2412-0

7. Index

FSC
www.fsc.org
MIX
Papier aus verantwortungsvollen Quellen
Paper from responsible sources
FSC® C105338